Anonymous

The Railroad System of California

Oakland and Vicinity

Anonymous

The Railroad System of California
Oakland and Vicinity

ISBN/EAN: 9783337417772

Printed in Europe, USA, Canada, Australia, Japan

Cover: Foto ©berggeist007 / pixelio.de

More available books at **www.hansebooks.com**

INTRODUCTORY.

Information respecting Oakland and its environs will be interestin███████
not only in California, but abroad. The terminus of a great system of railro██,
in rapid course of construction, must command the attention of all who are now,
or prospectively, interested in the commerce, or in the securities, of the tributary
roads ; and this, being the seat of numerous educational establishments, including
the State University, parents whose children are here, and those looking forward
to sending their children here, will feel interested in acquiring information concerning the place and its surroundings. Thus, the calculating merchant, the
shrewd operator, and thousands of thoughtful parents, will be gratified alike in
perusing the following pages.

JOHN B. FELTON,	ALEXANDER CAMPBELL,	J. WEST MARTIN,
BENJAMIN HAYNES,	HIRAM TUBBS, ·	GEO. C. POTTER,
JAMES B. LA RUE,	J. B. SCOTCHLER,	J. S. EMERY.

Publication Committee.

OAKLAND, October, 1871.

THE

▮ILROAD SYSTEM

OF

CALIFORNIA.

▮KLAND AND VICINITY.

▮TATE UNIVERSITY, ETC.

SAN FRANCISCO:

J. H. CARMANY & CO., "OVERLAND MONTHLY" PRINTING HOUSE, 409 WASHINGTON ST.

1871.

THE RAILROAD SYSTEM OF CALIFORNIA.

We publish, herewith, a reliable and interesting map of the railroad system of California which is concentrating at Oakland; also, a map of Oakland and its environs. In the following pages the reader will find a full explanation of both.

THE TWO GREAT COMPANIES.

The Central Pacific and the Southern Pacific now own, or control, all the railroads and railroad routes which are shown on the map, except the Stockton and Copperopolis, and the projected line from San Diego to Fort Mohave. Although distinct organizations, the affairs of the two companies are controlled by the same men. The concentration of the railroad system at Oakland may be regarded as a fixed fact.

NATURE HAS ORDAINED IT.

The trunk lines in California which have been subsidized by Government, were projected with more reference to subsidies than to the economy of railroad commerce. Of this, the Western and the Southern Pacific are notable instances. The Mt. Diablo range, extending south until it joins the Coast range, near San Luis Obispo, is the great obstacle to direct communication, by rail, between San Francisco and the interior, south of the 38th parallel.

THE LIVERMORE AND THE PANOCHE PASSES

Are, respectively, the routes of the Western Pacific and the Southern Pacific. The altitude of the former is 686 feet; that of the latter is 2,200 feet; and these are the most available passes in the range. Every 20 feet rise in a railroad grade is equivalent to an additional mile of level track; and every 360 degrees curvature is equivalent to a loss of half a mile.* Hence, the distance from Banta† to Niles‡

. * EQUATING FOR, GRADES.—The result of experiments, carefully conducted, gives as the resistance to motion of one ton, moving on a well-built level railroad, about 8½ pounds, or the weight which, hanging freely over a pulley, will overcome the friction of one ton. This resistance to motion is a constant fraction of the weight moved, and is its 1-264th part. This is the friction of the load. If, now, the plane be elevated from a level to a rise of 1-264th its length, according to well-known mechanical laws, one pound will, on this plane, sustain 264 pounds, or 8½ pounds will sustain one ton; and the fraction, 1-264, representing a rise of 20 feet in a mile, it follows that on this grade the effect of gravity is equal to that of friction, and in order to produce motion up this grade, *twice* the power must be applied that would be required were it on a level; and as it is a well-known mechanical law that the same amount of power is expended in raising a weight through a given height, whatever may be the *angle* of the plane upon which the motion is effected, it follows that for (see map and table of distances) being 44 miles, we must add 34 miles to arrive at its equivalent in level track, which would be 78 miles. We leave the loss by curvatures out of the question at present. After experiencing the enormous expense of running this part of the road, and keeping it in repair, the Company is about constructing a branch from Banta, to run *around* the mountains, via Antioch and Martinez, to Oakland. By this route trains can be brought

FROM THE SAN JOAQUIN VALLEY TO OAKLAND ON A LEVEL ROAD.

The distance from Banta to Antioch, over a level plain, is 30 miles; from Antioch to Martinez, via the shore, 18 miles; from Martinez to Oakland, via shore and valley, 31 miles — making, in all, 79 miles, with no curvatures of consequence. Now, considering the loss by curvatures and grades, on the mountain roads; the expense of keeping extra engines constantly on duty; the excessive wear and tear, etc., etc.—the reader can understand what he would not suspect from merely examining the map, to wit: that passengers and freight from the San Joaquin Valley, and from all the country east of Banta, tributary to the Western Pacific, will be transported from Banta to Oakland, via Martinez, much quicker, and very much cheaper, than it is possible to transport them from Banta to Niles—a point 22 miles south from Oakland, and 40 miles from Mission Bay, San Francisco, even by the much talked of route via Ravenswood and the "shore line."

The obstacles to be encountered on the proposed line of the Southern Pacific in crossing the Mt. Diablo Range, via Panoche Pass, are much more serious than those in the Livermore crossing. In the fifty miles breadth of mountains between Hollister and the San Joaquin Valley, the sinuosities of the road will be unprecedented, and the elevation to be overcome (2,200 feet) will add an equivalent of 110 miles, as compared with a level road. It is practically nearer from the Junction (north of Tulare Lake), via the valley route, to Banta, than from the Junction to the summit of the Panoche Pass; and the Southern Pacific trains, starting from the Junction, will reach Oakland, via the valley route, Banta, and Martinez, sooner than it will be possible for them to reach Gilroy, via the mountain road.

---

every 20 feet in height that we ascend on a railroad, we expend an amount of power equivalent to the transport of that weight over one mile of level; and this holds true whatever the grade may be. Equating for grades with a view to a comparison of lines, then, consists in *adding to the measured distance one mile for each and every twenty feet of ascent on the respective routes.*—APPLETON'S ENCYCLOPEDIA OF MECHANICS.

EQUATING FOR CURVES. — Direct motions on levels or inclines are affected less by disturbing causes than motion in curves; for, in addition to the irregularities growing out of the imperfections of the curved track and the varying elements of the curved motion in practice, are to be added all the disturbing causes which exist in the first case. This has, as yet, prevented that rigorous solution of the latter problem, which is to be desired, and which is essential to a true comparison, *a priori*, of the cost of movement on curved roads. It is, as yet, entirely an empirical formula deduced from a few experiments, but has been used for the purpose of comparison of routes by distinguished engineers, and is the best we can offer with our present knowledge of the subject. We find by the experiments referred to above, that a curve of 400 feet radius *doubles* the resistance. In propelling a train, then, through an entire circumference of such a curve, we expend twice the power that would be consumed in traveling an equal distance in a right line. Taking, then, the analogy afforded by motion on ascents as compared with levels as a guide, and we conclude that the same power would be expended in turning through an entire circle, *whatever may be its radius*, (this, of course, must be understood as confined to certain limits); hence, for every circle of 360 degrees, we must add for the expenditure of power on a right line of the same length, the circumference of a circle described with the radius of double resistance, found by experiment as above to be 400 feet; this will be half a mile. Equating for curves consists, then, in *adding to the measured distance one-half mile for each and every three hundred and sixty degrees of curvature on the respective routes.*—APPLETON'S ENCYCLOPEDIA OF MECHANICS.

† Banta is on the line of the Western Pacific, three miles west of the San Joaquin River.

‡ Niles is on the Alameda Plain, opposite to the gap where the railroad enters the hills.

According to General Palmer's report, grades will be encountered in the
Panoche Pass where it will require four or five extra engines to perform what one
engine will accomplish on the valley road. Thus, in railroading, it is true that a
pot-handle is longer standing up than lying down; and that the longest way round
is often the shortest way home.

Enough has been said to demonstrate that the great passenger and freight
route to Oakland and San Francisco will

APPROACH OAKLAND FROM THE NORTH.

The road from Martinez to Oakland will also constitute the connecting link of the
"air-line road" between San Francisco and the Central Pacific overland road, via
Sacramento; between San Francisco and the road to Puget's Sound, traversing a
continuous line of productive valleys in California, Oregon, and Washington; it will
connect with the projected road which will tap the Sacramento Valley west of the
river, and extend to Red Bluff; it will unite with the system of roads which are
tapping Napa, Sonoma, Petaluma, Santa Rosa, and the Russian River Valleys.

THE LINE APPROACHING US FROM THE SOUTH,

Even though it can not successfully tap the San Joaquin Valley, is yet of much
importance. The Santa Clara Valley, alone, embraces a vast area of highly productive land, and is capable of supporting a larger population than some of the New
England States. Extending southerly to Santa Barbara is a succession of smaller
valleys, to which we will hereafter refer. That section of the Southern Pacific
road west of the Panoche Pass, in connection with the projected Coast Line, will,
therefore, do a large business, both in passengers and freight. San José — the
point of divergence of the roads running north — is ten miles nearer Oakland than
San Francisco. The Oakland road is level, and the San Francisco road has
severe grades. Supposing, however, the places were equidistant, and the roads
equally level, the export products would seek Oakland, where the economy of
shipping exceeds the possibilities at San Francisco.

TERMINAL FACILITIES.

The Central Pacific Railroad Company has secured, in and about Oakland,
facilities for the conjunction of railroad and water traffic unequaled in the world,
and unattainable elsewhere on the Pacific Coast. It owns, in fee simple forever,
an area of seventy acres fronting on the Bay, in the western part of the city, which
it purchased as a site for machine-shops, etc. It also owns an area of three hundred and fifty acres on the water front, extending from the former tract toward
Goat Island, with a frontage of nearly half a mile on ship channel. It also owns
extensive reservations on the southern bank of the estuary of San Antonio, and
it has secured the right of way for tracks leading to them from the main trunk.
It also owns, in proximity to Oakland, ninety acres of land suitable for a car-yard
and other uses; and a large tract of hilly ground whence it can obtain, *ad libitum*,
earth and gravel for filling purposes.

The improvements at ship channel are described elsewhere in an article taken
from the *Alta;* but instead of being *completed,* as the *Alta* presumed, they constitute a small part of a grand design. The Company is exempt from the operation of State and municipal laws respecting wharfage, dockage, and tolls; and it
imposes no charges upon vessels receiving or delivering freight.

Infinite credit is due Messrs. Stanford & Co. for having thus early secured the estate and franchises which will afford such invaluable and unbounded terminal facilities; and it is a subject for congratulation to the people of the coast at large that, while the railroad system is developing industrial resources with unexampled rapidity, means are assured for the most economical handling of exports and imports.   The reflection that the most productive farming, pasture, vineyard, and orchard lands of California and Oregon are being placed in direct communication with ships floating in the waters of the Pacific, and with the entire railroad system of the United States, is suggestive of an export commerce that will soon turn the balance of trade in our favor, and keep more of our gold at home.

THE CENTRAL PACIFIC

Track commences on the Oakland water front at 26½ feet of water, (at low tide), where Goat Island makes a lee in "north-westers," and the Alameda Encinal breaks the force of "south-easters."   The main trunk runs thence, southerly, to Vallejo's Mill (see map), from which point it runs eastwardly through Livermore Pass, traversing the Suñol, Amador, Livermore, San Joaquin, and Sacramento Valleys, to the Sierra Nevada, passing through Stockton and Sacramento on the route across the continent.   A branch is in operation southerly from Vallejo's Mill to San José, connecting with the line of the Southern Pacific which is now built to Gilroy, and is being extended southerly toward the Panoche Pass.   We will omit descriptive details of the road and route, inasmuch as the public are familiarized with both.

THE CENTRAL PACIFIC SAN JOAQUIN BRANCH

Is one of the most important feeders of the Central Pacific main trunk.   It intersects the main trunk eight miles westerly from Stockton, and runs southerly forty miles, through a portion of the great San Joaquin Valley, surnamed "Paradise"— one of the most thickly settled agricultural districts in California.   In proportion to its length, it is, perhaps, the most valuable "feeder" which the Company could have constructed.   It is now being extended to meet the agricultural developments of the San Joaquin Valley, and will eventually intersect the trunk line of the Southern Pacific, in the neighborhood of Tulare Lake.   Thus, the empire valley of the Pacific Coast is destined to be traversed by two great roads; and the time will come when both will require numerous branches, to accommodate the vast breadth of arable country which the valley embraces.

THE BRANCH FROM BANTA, VIA ANTIOCH AND MARTINEZ, TO OAKLAND,

Has already been described, but it is destined to work so important a part in our railroad system, that we recur to it, under a special heading.   It is the solution of a State problem.   Our most extensive wheat-districts, and our coal-mines, *will have a level road to the sea.*

THE CALIFORNIA AND OREGON,

This road, which is to concentrate the trade of the north and bring it to the Bay of San Francisco, is being constructed by the Central Pacific Railroad Company.   A valuable land grant is stimulating the work.   It is now completed from Sacramento to Tehama—123 miles.   While a full force is being employed in its northward extension, the Company is preparing to make the connection between Sacramento and Oakland by the shortest possible route.   The engineers have found a route of eighty-three miles, on which no grade will be encountered exceeding ten feet to the mile. It crosses the straits in the neighborhood of Martinez, where it will intersect the road from Banta to Oakland.

It is remarkable that neither the press nor the people of San Francisco have manifested the slightest solicitude for the railroad trade from the north, even when there was danger of losing it, while they have been subject to intermittent fever concerning that from the south, of which they have always been assured, but which is neither now, nor prospectively, half as important as the former.

THE CALIFORNIA PACIFIC, AND THE N. S. F. AND HUMBOLDT

Comprise the road extending from Vallejo to Marysville, with a branch to Sacramento; the road from Calistoga to Napa; and the system of roads, and projected roads in Sonoma, Petaluma, Santa Rosa, and Russian River Valleys. Prior to the purchase of these lines by the Central Pacific, Vallejo promised to be an important terminus, and Donohue had no small pretensions. Both places are likely to remain important points for local trade; but when the California and Oregon Road makes its connection with Oakland, the concentrating tendency of commerce will be illustrated for the ten-thousandth time.

THE WOODLAND AND RED BLUFF BRANCH

Which will traverse the richest part of the Sacramento Valley, west of the river, for a hundred miles, is a projected road, much needed, and one that will assuredly be constructed in a few years. There is no valley in the State that would yield quicker or richer returns to a railroad.

THE SOUTHERN PACIFIC.

The original franchise of this Company extended "from the waters of the Bay of San Francisco to a point on the Colorado River, at or near Fort Mohave," where it meets that of the Atlantic and Pacific. It has since obtained from Congress a land-grant for a branch from the Tehachepa Pass, via Los Angeles, to Fort Yuma where it will intersect the line of the Memphis and El Paso. The topography of the country does not admit of a more direct route.

The tardiness in the prosecution of the work is attributable to the great difficulties encountered in crossing the Mt. Diablo Range. The Livermore Pass having been secured by the Western Pacific, the Southern Pacific was given the option of taking any one of four passes farther south, to wit: the Pacheco, Panoche, San Benito, or Cholame.* Much time, and immense labor, have been expended in seeking the most available route.

The report of General William Palmer on surveys across the continent, on the 35th parallel of latitude, published in 1868, throws light on this interesting subject.

The surveys were begun at Fort Wallace, in western Kansas, in July, 1867, by three well-organized parties of engineers, under General W. W. Wright. Two additional parties, under Colonel William H. Greenwood, were subsequently sent out, increasing the corps to five parties comprising about one hundred men, besides the military escorts, teamsters, etc. The work was thorough and exhaustive. It extended over the mountainous regions and arid plains, and the contingencies of climate and seasons were investigated.

---

* On the map filed with the Secretary of the Interior, the route takes the San Benito; but an Act of Congress allows the Company to cross the mountains within thirty miles on either side, "or as near as may be."

The line recommended for reaching the Tulare Valley from the east, crosses the Colorado River about twenty-five miles below Fort Mohave, and traverses the Mohave Plains to the eastern foot of the Sierra Nevada. The Tehachepa Pass, about forty miles east and north of Tejon, was found to be the best at which to cross that great range. The elevation of the summit is 4,008* feet above tide, while at the Donner Lake Pass, where the Central Pacific Railroad crosses the same range, the altitude exceeds 7,000 feet.

Descending from the Tehachepa into the Tulare and San Joaquin Valleys, General Palmer sought a route through the Mt. Diablo Range by one of the four passes before-mentioned. Of these, the Panoche was the only pass instrumentally examined, the elevation of which was found to be about 2,200 feet above tide. The grades in 38 miles, from Tulare Plain across the Coast range, are as follows: 7 miles, of 106 feet per mile, ascending westward; 6 miles, of 116 feet per mile, descending westward; and the remaining 25 miles ranging from 50 to 85 feet. (General Palmer's Report, p. 71.)

· The elevations of the three other passes, as ascertained by the barometer, are as follow: Pacheco, 1,470 feet; San Benito, 2,750 feet; Cholame, 2,000 feet. The lowest, Pacheco, is described by General Palmer as being the most difficult of all. A peculiarity of the whole range, is the abruptness of the slopes from the Summit to the San Joaquin and Tulare Valleys. All the passes are easily approached from the westward; but steep, and in some cases impracticable, grades are required to make the descent into the valley. The sand formation of the country is also exceedingly unfavorable for the construction and maintenance of railroads.

The vast amount of subsequent surveying has failed to develop a more favorable route for reaching San Francisco from the south, than the one recommended by General· Palmer.

While the subsidy amply justifies the building of the road through Panoche Pass,

THE VALLEY ROUTE

Which will traverse the level plain (the average gradient being less than two feet to the mile) on an air line from the Junction to Banta, connecting with the branch road from Banta, via Martinez, to Oakland, will be—as elsewhere explained— much the quicker and cheaper approach to San Francisco from the Tehachepa Pass, and from every part of the Tulare and San Joaquin Valleys.

The country tributary to the mountain route is comparatively unproductive, while that tributary to the valley route is remarkable for its immense agricultural resources.

THE SOUTHERN PACIFIC COAST ROUTE,

As shown on the map—extending from Gilroy to Los Angeles—was projected by General Rosecrans, and originally designed to connect with the Memphis and El Paso Road, at Fort Yuma. As its name indicates, it now belongs to the Southern Pacific Railroad Company. The route traverses a chain of valleys from Gilroy to Santa Barbara which, though not comparable in extent to the valleys farther inland, are remarkable for salubrity of climate and fertility of soil. From Santa Barbara to Los Angeles, the country is rough and broken, presenting serious engineering difficulties. The building of the Memphis and El Paso Road would stimulate

---

* The great breadth of the range renders the grades comparatively easy.

the building of this projected coast road, and cause its extension beyond Los Angeles to Fort Yuma, thus making the connection with Oakland and San Francisco. The old proverb, "All roads lead to Rome," has a significant application to our railroad system—terminal expenses and transshipping facilities rendering it an economical necessity.

THE MEMPHIS AND EL PASO.

This enterprise was one of the first of the Pacific Railroad projects presented to the public, and it has been prosecuted with varying fortunes for sixteen years. The Company was organized in 1854, and received a valuable land grant from the State of Texas. Work was soon after commenced, and about two hundred and fifty miles of the road have been graded and put in operation. The late civil war caused a suspension of the work, and the exhaustion of the South has, until quite recently, prevented its resumption. Last winter, a bill passed Congress granting a subsidy to encourage the building of the road.

THE STOCKTON AND COPPEROPOLIS

Is a local road, designed to connect Stockton and Copperopolis. Fifteen miles of it has been constructed, and the road is in operation that far, easterly from Stockton.

[From the San Francisco Alta California.]

WHAT THE C. P. R. R. CO. HAS BEEN DOING.

A wharf, 11,000 feet long, running out to a depth of 26½ feet at low tide, and 33½ feet at high tide, in a bay like that of San Francisco, having twelve railroad tracks upon its last one thousand feet, a wide carriage way, a spacious passenger depot and railroad offices, warehouses and outside storage for 40,000 tons of grain or other merchandise, three large docks, one of which affords ample space for five of the largest steamers or clippers afloat, is not often seen, even in this age of railroad and engineering wonders. Such a structure has, however, recently been completed by the Central Pacific Railroad Company on the Oakland or easterly side of the Bay of San Francisco. The extreme end of the main wharf is only three miles from the foot of Second Street, where freight is landed in this city, and is less than two and a half miles from the foot of Pacific Street, where passengers are set down on this (San Francisco) side.

It would be much more difficult to build such a long wharf with safety on our side of the bay, because the bottom here is a yielding mud; but on the Oakland side there is a hard clay bottom. Another point in favor of Oakland is that the destructive marine wood-eating worm, the *teredo*, is not found there at all. In these facts lie two of the greatest elements of strength and ability to bear great burdens of the new railway wharves across the bay, but only two of them. Nothing has been neglected in the quality of material used, and workmanship employed, to make the wharves the very best ever built in the United States. Experts in the construction of such work, and army and other engineers who are familiar with like structures in Europe and America, all agree in saying that for engineering skill, mechanical execution, and solidity and excellence of all the material employed, the work is not surpassed by any they ever examined. To make this plain, a few descriptive figures must be given. The piles used, where the water deepens, are 65 feet long, and are 42 to 54 inches in circumference, or as heavy as the main-mast of the largest clipper. They are all of the very best pine, which for lasting qualities under water is one of the very best kinds of wood. The main wharf—for a thousand feet east of the latter there are two wharves, one for Oakland local trains, and one for the regular freight and passenger cars of the through overland road—is 800 feet wide at the extreme or western end, and on it are pens for 500 cattle, two immense warehouses (one 50 x 500, another 62 x 600), with the passenger depot, 75 x 305 in size. The piles were driven into the bottom to a depth of 18 feet. They are set 10 feet apart, parallel with the course of the wharf, and 6 and 7 feet apart across it. In the docks, or slips, there is a double row of spring or fend-off piles, and the regularity and neatness with which they are laid is especially worthy of attention and admiration. Those who will examine the old slips into which the steamers used to run, or who remember those used at the Brooklyn (New York) ferries, will be able to appreciate the superiority of the Central Pacific slips. The upright piles on the last one thousand feet of the main wharf, are braced with immense cross piles and iron anchors. Trains of heavily loaded freight cars pass over this gigantic structure with as little jar as over solid ground.

The massive new freight ferry-boat of the Central Pacific Company has been completed, and is now running from the Company's extensive wharves, at the foot of Second Street, in this city, to the Company's wharves and docks above described, on the Oakland side of the bay. The boat carries 16 loaded cars on each trip, and has, in addition, pens for 300 cattle. She can carry from 1,000 to 1,280 tons each way per day, making the trip across the bay (3 miles) in forty minutes, when loaded. A railroad now connects the Pacific Mail Company's dock and the Central Pacific wharf on this side, by means of which the cargoes of the China steamers can be immediately discharged into the Central Pacific freight cars, and thus the utmost dispatch will be attained in the shipment of teas, silks, and other fast freight intended for the Atlantic States and Europe.

The Central Pacific Company owns all of the Oakland water-front on which its vast wharves are built. It has its own docks there, capable, as we have shown, of accommodating five of the largest clippers or steamers at a time. In future, all grain, ores, wool, wine, and other merchandise that are to be shipped to Europe or elsewhere, by water, will be discharged directly on shipboard from the cars at the end of the Company's wharf, while all steamers or other craft which come into this port with Oriental cargoes for the Atlantic States, will go direct to the Company's docks and unload into the cars. In this way, heavy wharfage, tolls, loss of time, double handling and its onerous attendant costs, will be avoided.

In addition to the main overland line, the Central Pacific Company owns the California and Oregon Railroad, which is now completed to Tehama, 123 miles above Sacramento, toward Oregon ; the San Joaquin Valley Road, which is completed to the Tuolumne River ; the San José branch, from Niles' Station, and the Alameda and Oakland Railroad. The two first-named branches of the Central Pacific line run through our two great valleys—the Sacramento and San Joaquin.

No country on earth offers a more princely traffic to railroads than do these two vast empire valleys.

In the building of these deep-water wharves and warehouses, the Central Pacific Company has omitted nothing which would tend to facilitate business and reduce expenses on the Company's railroads. Mr. S. S. Montague, the Chief Engineer of the Central Pacific Road, and Mr. Arthur Brown, who built all the Howe truss-bridges on the road, planned and built the great wharves and warehouses which we have described, and the whole work stands as a fresh monument of the engineering talent of the one, and the mechanical skill and ability of the other.*

CONCLUSION.

Railroad companies conserve their own interests best, when they promote the interests of the public. The Central, and the Southern Pacific, in seeking the patronage and sympathetic coöperation of the populations for whose necessities they intend to provide, will operate with the greatest possible economy to the public, and to themselves. To do this, they must seek the nearest point at deep water, convenient to the sea, by such routes as are, equivalently, the shortest and cheapest. Oakland is that point. The Mt. Diablo Range must be removed, or split asunder, before the figures we have quoted, in equating for grades and curves, can be controverted. And what do they demonstrate?

First, that even if the Bay were bridged at Ravenswood, and a shore-line road extended, thence, to Mission Bay, trains leaving Banta for Oakland, via Martinez, would reach Oakland before similar trains leaving Banta, at the same time, for San Francisco, via Livermore Pass and Ravenswood, could get within *forty miles* of San Francisco.

Second, that the Southern Pacific trains, starting from a given point in the Tulare Valley, will reach Oakland, by the valley route, before said trains could get within *eighty miles* of San Francisco, via the mountain road.

Third, that a bridge at Ravenswood would be, to San Francisco, a bridge of *size.*

DISTANCES.

In columns A, B, and E, of the following table, the measured distances are given, except in cases wherein they have not been made public. In these cases they have been computed by engineers who are familiar with the general topography of the country. In columns C, and D, the respective elevations of the Livermore and Panoche Passes have been taken into account; and in conformity with the established rules in equating for grades, (see note, page 3), 34 miles have been added to the measured distance through the former pass, and 110 miles to the computed distance through the latter, to compensate for grades.

This gives the reader an intelligent idea of the equivalent, or practical, distances, via the several routes, relative to the *power* required for transportation by rail, in

---

* While according infinite praise to Messrs. Montague and Brown for their genius in designing, no less praise is due to Mr. A. R. Guppy, the accomplished and indefatigable engineer who directed and superintended the work. We will add that the work done is only a small part of that which is projected.

ascending grades. It does not, however, impart what should be understood respecting the *time* consumed on steep grades, as compared with level road. To prevent the "iron horse" from running away with the train, in descending such grades, it is necessary to "down brakes," and "go slow." Thus, generally, the descent requires as much time as the ascent, and the rules in equating for *speed* tell heavily against mountain roads. Nor, has the loss by curvature been estimated, either in the foregoing remarks, or in the following table—the loss by grades being amply sufficient to sustain all that is claimed in the text.

It is apparent that the level road to the sea, which will run *around* the mountains, and approach Oakland from the north, must become the great trunk line of both the Central, and Southern Pacific.

The distance from Oakland to Martinez is computed, as will be observed, at 31 miles. There is reason for believing that the railroad company has located a line between the two places that will not exceed 26 miles in length; but the former figures are adopted in the table, as the maximum.

FROM OAKLAND TO	A Via Boat and Vallejo Route.....	B Via Martinez, Antioch, and Banta....	C Via Livermore Pass Route.....	D Via San José, Gilroy, and Panoche Pass.	E Via Air-line Route...
Niles (near Vallejo's Mill).....	.....	.....	22	.....	.....
Martinez.....	37	31	.....	.....	.....
Banta.....	.....	79	100	.....	.....
Stockton.....	.....	95	116	.....	.....
Sacramento.....	91	.....	164	.....	83
Vallejo.....	31	.....	.....	.....	26
Napa.....	47	.....	.....	.....	42
Calistoga.....	74	.....	.....	.....	69
Sonoma.....	55	.....	.....	.....	50
Santa Rosa.....	79	.....	.....	.....	74
Healdsburg.....	93	.....	.....	.....	88
Woodland.....	86½	.....	.....	.....	80
Marysville.....	119	.....	192	.....	108
Red Bluff.....	210	279	300	.....	210
Portland (Oregon).....	704	739	760	.....	.....
San José.....	.....	.....	.....	40	.....
Gilroy.....	.....	.....	.....	70	.....
Hollister.....	.....	.....	.....	83	.....
Junction (north of Tulare Lake).....	.....	227	248	310	.....
Tehachepa Pass.....	.....	327	348	410	.....

The distance from Niles in a direct line across the Bay to Ravenswood is 13 miles; thence to Mission Bay, San Francisco, via "shore line," 27 miles—total distance from Niles to San Francisco, 40 miles, as against 22 miles from Niles to Oakland. San José is 50 miles from San Francisco, and but 40 miles from Oakland.

OAKLAND AND VICINITY.

THE CITY GOVERNMENT.

MAYOR, ... N. W. Spaulding.

CITY COUNCIL — E. H. Pardee, President; J. V. B. Goodrich, T. J. Murphy, A. L. Warner, C. D. Haven, W. J. Gurnett, W. S. Snook.

BOARD OF EDUCATION — L. Hamilton, President; E. W. Playter, G. W. Armes, R. E. Cole, Jacob Bacon, W. Van Dyke, J. W. Thurman.

CITY CLERK AND TREASURER, ... H. Hillebrand.

CITY MARSHAL AND TAX COLLECTOR, . . . Perry Johnson.

POLICE JUDGE, ... A. H. Jayne.

POLICE COMMISSIONERS—N. W. Spaulding, E. H. Pardee, and Perry Johnson.

CITY ASSESSOR, ... Joseph M. Dillon.

JUSTICES OF THE PEACE, ... James Lentell and G. H. Fogg.

SUPERINTENDENT OF PUBLIC SCHOOLS, ... F. M. Campbell.

APPOINTED OFFICERS — T. J. Arnold, City Engineer; H. H. Havens, City Attorney; George Taylor, Pound Master; Miles Doody, Chief Engineer of Fire Department.

POLICE DEPARTMENT — F. B. Tarbett, Captain of Police; D. H. Rand and E. H. Woolsey, Detectives; W. P. Brandt, G. H. Moore. W. H. Summers, John A. Moore, Spencer Pool, H. C. Emmons, George H. Carlton, C. P. McKay, Regular Officers; A. Shorey, A. Wilson, G. F. Blake, G. H. Tilly, Special Officers.

[From the Oakland Transcript.]

THE PAST, PRESENT, AND FUTURE OF OAKLAND.

The centralization of society, the development of industries, and the current of trade, being subjects of general interest, the following synoptic review and brief deductions concerning the locality of Oakland are appropriate at the present time.

Before Oakland existed, San Francisco had become the great centre of population and trade on the North Pacific Coast. Admirably situated for deep-sea and inland water traffic, wealth was attracted to her lap. This stimulated the enterprise of her people, and made her what she is. Sacramento, Stockton, San José, Benicia, Vallejo, Sonoma, Petaluma, (to say nothing of numerous mountain towns, which dot the map of California), all acquired considerable importance before Oakland was heard of.

On New Year's Day, 1851, the site of Oakland was known only as a part of the Peralta Rancho. Wild cattle roamed where now, surrounded by all that pertains to modern civilization, more than eleven thousand people are living. The sound of church-organs and college-bells now reverberates where, then, nothing but the bellowing of animals interrupted the stillness of nature. In the place of the old cattle-trails are railroads and macadamized streets; and where the cattle lazily roamed, we now witness thirty-two daily passenger trains, to say nothing of freight trains, rushing to and fro, propelled by the mighty power of steam. Even the wild flowers, that once bedecked the surface of the earth, exist only by sufferance, and a cultivated flora has usurped their place.

Considering that Oakland was but a *second thought* in California; considering the long litigation concerning land titles—now happily settled; considering that one-fourth the area of the city has been held in check for the want of public thoroughfares—the circumstance of her extraordinary development, the statistics of which we publish elsewhere, affords a useful lesson for economists.

Our space is inadequate to a full exposition of the subject, but we will dwell upon it sufficiently to explain "the milk in the cocoanut;" and to show that more extraordinary results will inevitably succeed those which it has been our privilege to witness.

For several years after the acquisition of California by the United States, men "planted their stakes" on the exclusive basis of the gold and silver crop, and the trade which mining would develop. Moreover, in their calculations concerning prospective developments, ships, steamboats, and mule-teams were relied upon as the only means of transportation. In short, a single branch of industry was the incentive to action, and the Locomotive was not even *expected* within the time popularly allotted for making "a pile."

The Locomotive has not only revolutionized the carrying trade, but, while adding importance to mining industry, it has stimulated agriculture to the front rank, and opened many fields for diversified labor. The gold and silver crop can be "packed" from the mountains to the sea on the backs of mules, and requires not much tonnage to transport it from continent to continent; but the wheat, wine, wool, and fruit crops will annually require hundreds of vessels and thousands of railroad trains.

The statistics, and our remarks elsewhere, will show what the Locomotive has thus far done for Oakland, in connection with educational establishments, and natural advantages of climate, soil, and topography. Respecting the present, we will only say, here, that there is no other city, in or out of California, the population of which includes so large a proportion of the well-educated class.

Referring to the changes produced, and being produced, by railroads, the unbiased reader need only examine the map to see that there can be no great terminus at ship-channel in the Bay of San Francisco, except at Oakland. An "airline road," so called, will soon be made from Sacramento to Oakland, and engineers are in the field to determine the shortest route.

Plans are almost completed for dredging the bar at the mouth of San Antonio Estuary, and making the estuary available for commercial purposes. An important consideration, in connection with the vast amount of piling already done, and the vaster amount in contemplation, is the absence of the *teredo*, or "pile-worm."

Scientific men attribute this to the fact that the flood-tide through Raccoon Straits throws the fresh water from the Sacramento and San Joaquin, far over to the east side of the Bay; and the pressure of the flood, south of Angel Island, crowds it sufficiently to cause a portion of it to pass between Oakland and Goat Island, at every ebb. The absence of the *teredo* from the estuary has never been accounted for satisfactorily, but the fact of its non-existence is established.

We have written enough to show that Oakland must eventually become the base of the greater part of the commerce concentrating at the Bay of San Francisco. The situation of Oakland toward San Francisco, is often compared with the situation of Brooklyn toward New York, and comparative deductions are made corresponding with the history of those Eastern cities. Had New York been located at the end of a peninsula, jutting from the main-land into the Atlantic Ocean; and had Brooklyn been located on the main-land opposite, and enjoyed a climate as much more genial as that of Oakland, compared with the climate of San Francisco, we opine the result, there, would have been different.

In writing thus about Oakland, it must not be supposed we are predicting the downfall of San Francisco. On the contrary, we believe that San Francisco will prosper and increase. We are looking to the time when the commerce concentrating at the Bay of San Francisco will be fivefold greater than at present. And without expecting Oakland to depopulate her great neighbor, we judge, from the forces which are operating, that our next annual statistics will make a more wonderful showing than those of the past year.

THE WATER FRONT.

What is known here as "The Water Front of Oakland," consists of the tide-lands embraced within the charter line of the city, as shown on the map published herewith. This does not refer to the marsh-lands—they being above ordinary, or average, high tide. Some characteristics of this water front are remarkable. The bed of the San Antonio Estuary, and of its main current to ship-channel, is soft. and offers a great reward, in commercial value, for engineering skill. The flat, from the shore of the bay to ship-channel, dips from high-water mark, westerly, on a regular incline. It is "hard-pan," and presents an admirable foundation for wharves and other structures.

By the Act incorporating the town of Oakland, passed 1852, the State ceded the water front to the town. By a subsequent Act, the town became a city, and the old charter line was confirmed. In 1852, the Trustees of the town conveyed the entire water front to an individual, for a consideration—*such as it was.* The city authorities repudiated the action of the Town Trustees, and sought to recover the water front. A prolonged litigation ensued, the merits of which belong to the past. However much the development of the city was retarded, some of the results have been interesting. One, was the preservation of the water front in its integrity, as a whole; and when the transcontinental railroad sought its natural terminus at the Bay of San Francisco, the Genius of Commerce was invoked, and she extended an open hand. The city sought, and obtained from the Legislature an "enabling Act," under the provisions of which the litigation was concluded. and her claims to the water front were exchanged for guarantees of metropolitan portensions. Master minds were employed; grand conceptions were developed; great things have been accomplished; and greater things are in progress.

The water front, excepting the portion of it owned by the railroad company, and a small reservation belonging to the city, is the property of an incorporated company, known as The Oakland Water Front Company, in which the directors of the railroad company are largely interested. An elaborate survey of the entire property has been completed, and the contemplated improvements, (an outline of which we publish), as shown on the Company's map, develops one of the greatest enterprises of this or any other age.

THE ESTUARY OF SAN ANTONIO.

An examination of the maps published in this pamphlet will convey a better general idea of the location of the Estuary, with reference to the Bay, the city of San Francisco, Oakland, the railroad system of the Pacific Coast, and the commerce of the ocean, than we could convey in words.

On the map of Oakland will be found the outlines of the reservations and rights of way, on the south side of the Estuary, belonging to the Railroad Company; also, the outlines of the improvements projected by the Water Front Company, which harmonize with those of the former.

The reader will observe the soundings marked on the map, from four and a half fathoms water in the Bay, to the head of the Estuary; and the scale will enable him to judge of the area of this most invaluable, land-locked, sheet of water.

Our article upon the Water Front of Oakland explains the situation of the Estuary, in the relation of ownership; and the proposed line of crib-work, as shown on the map—extending from ship-channel, in the Bay, to the head of the Estuary—is the line established by the engineers employed by the Water Front Company, and has been copied, by permission, from an elaborate-map which the Company has prepared.

As we have stated, elsewhere, the plans of the Company develop one of the grandest conceptions of this, or any other age. Recognizing the immutability of the law of economy, it has comprehended the era of railroad commerce, and its relationship to the commerce of the ocean. It has formed a partnership with Nature where Nature furnishes nine-tenths of the capital.

The improvement of a portion of the river Clyde which is now contributary to one of the greatest centres of industry in the world, cost several millions of dollars; but the Estuary of San Antonio, with a capacity for thirteen miles of land-locked wharfing, and a basin to float a fleet of the largest vessels; which is in close alliance with the terminus of a continental railroad system; and on the banks of which, locomotives from New York, Philadelphia, and Boston; from Chicago, Cincinnati, and St. Louis; from New Orleans, Mobile, and Charleston; can stand in waiting to whistle greeting to steamers from Panama, Sydney, and Honolulu; from Astoria, Yokohama, and Japan—this Estuary can be made immensely contributary to the commerce of the world, at an expense of a few hundreds of thousands of dollars.

CLIMATE.

Information respecting climate being already widely disseminated, the reader will be more interested in general comparative remarks than in meteorological details.

Often, the thermometer is a poor index to the comfortable temperature in California. A degree of heat or cold that is not distressing in one locality, is almost insupportable in another. In the dry atmosphere of the mountains, ice forms in the shade, when nobody feels uncomfortably cold; and in the humid atmosphere of the sea-coast, ice melts in a blanket, when every body is chilled to the bone. When the mercury indicates a temperature of 80 degrees, people swelter in a humid atmosphere, and refrigerate in a dry one. Therefore, taking it for granted that information about bodily comfort will be more interesting than minutes concerning the range of the mercury, we shall devote more attention to the former than to the latter.

Taking the climate of San Francisco as a basis for comparison, the mean annual temperature for seventeen years, as determined by Dr. H. Gibbons, Sr., of that city, was 56° 4'—the mean temperature of spring having been 56° 5'; summer, 60°; autumn, 59°; and winter, 50°. There were but six days when the mercury reached as high as 90°, and but one day when it fell as low as 25°. During the wet season, the climate of the country surrounding the Bay varies little from that of San Francisco; but during the dry season the variations are remarkable.

The rarefaction of the air, produced by the action of the sun's rays upon the vast surface of the interior country, is the cause of our prevailing summer coast-winds. The air is drawn from the ocean to re-establish the equilibrium (inland) which is destroyed by the heat. The force of the wind depends on the degree of rarefaction that has been produced, and its direction is influenced by intervening obstacles presented by the topographical features of the country.

At some places, the wind and fogs from the ocean sweep over the surface; some places are protected from the force of the wind and the humidity of fogs by the configuration of the mountains, but are often deprived of the sun's rays by the fogs passing overhead; others are protected entirely from the wind, and enjoy an unclouded atmosphere which permits the accumulation of heat; and, again, the gravitating tendency of a cold current from the ocean often causes it to sweep down the lee slope of the hills, or to dip to the surface of the plain, between two ranges. Hence, the difference in the sensation of heat and cold experienced at places only a few miles apart. The necessity of substituting cloth wrappings for lawns or linen, within a transit of thirty minutes by boat or rail, seems wonderful, even when we know the cause.

The summer climate of Oakland and vicinity, is a matter of curiosity to many. Immediately back of Oakland, the mountains are high, but there are depressions in the range, both north and south of us, at a distance of several miles. The strongest wind-currents are, of course, drawn through these depressions. We see the fog banks which enter the Golden Gate take a northerly direction, and the fog banks which come through the "Mission Pass," in the southerly part of San Francisco, take a southerly direction, across the Bay. The high hills between the central part of San Francisco and the ocean often protect that portion of the city from a low fog bank; but, even when the fog bank is high, and envelops San Francisco in its humid embrace, Oakland almost invariably escapes it. When the fog bank is so dense and heavy that the depressions in the mountain range, north and south of us, do not accommodate it, and the fog from either direction meets overhead, it is generally absorbed, before reaching the earth, by the accumulation of dry, heated air; or lifted from the surface, before reaching Oakland, by the upward tendency of the draught which must pass *over* the high mountains behind us. Thus, the most important difference between the climate of San Francisco and Oakland, is attributable to the configuration of the neighboring mountains. The movement of the fog indicates the force and direction of the wind, and every boy who has ever sat on the windward side of a board fence, and enjoyed being out of the wind, will understand the foregoing explanation.

The difference in temperature between Oakland and San Francisco, as indicated by the thermometer, is not so great as many persons suppose; but the difference in the velocity of the wind and in the humidity of the atmosphere, is the chief cause of the contrast in comfort and health between the two places.

During the prevailing summer winds, our climate is a mean between that of San Francisco and San José. Winds from the north or north-west, which come in a direction nearly parallel with the Coast Range, are more violent at Oakland than at San Francisco; but they are of rare occurrence.

SOIL AND PRODUCTIONS.

The soil of Oakland is a sandy loam, varying from three to four feet deep. Beyond Oakland, toward the foot-hills, it partakes more of the pure loam, or *adobe.* In the northern part of the city (the part toward the foot-hills) it is less sandy than in other places. The apple, pear, plum, cherry, and apricot, are produced in great perfection wheresoever planted. The almond also thrives, and bears plentifully. All kinds of garden vegetables, except the egg-plant and okra, can be produced at will, and in great abundance. Raspberries, strawberries, and currants, thrive and bear marvelously. Shade and ornamental trees make rapid growth, as the gardens on every side attest. So much has been written about our productions that we were inclined to omit the subject. Indeed, the cultivation of fruits and vegetables has almost ceased in Oakland. Ornamental trees, shrubs, and flowers, are preferred. The nursery gardens in the vicinity afford an evidence of the public taste for the beautiful in Nature. For example, in the "Belle View Nursery" are found forty-two varieties of the acacia, thirty-three of eucalyptus, ten known varieties of California oak, and more than one hundred varieties of coniferæ, to say nothing of thousands of shrubs and tens of thousands of flowers.

As a rule, we can gather beautiful bouquets from plants in the open air every month in the year. In sheltered situations, the fuchsia, oleander, geranium, and even the heliotrope, withstand our severest winters.

THE NATURAL SUPPLY OF WATER.

In every part of Oakland water can be obtained from wells ranging in depth from 14 feet to 35 feet. Taking the neighborhood of Eighth and Center streets as the mean, we find two wells, eight feet eight inches in diameter, and twenty-five feet deep, which yield, when the water is lowest, all that can be pumped by a single-horse-power, working ten hours per day. The proprietor of one, informs us that, at times, his well has been taxed at the rate of 10,000 gallons per day. Each of these wells has preserved the verdure of extensive lawns during the past summer, (the driest we have experienced), and the water in both is soft and pure. A corresponding supply of water is obtainable in every part of Oakland, from wells of the same diameter; but the requisite depth of wells depends on the profile of the ground, and varies as before mentioned.

The force of the wind, although not so uniform, nor so great, in this neighborhood as at San Francisco, is amply sufficient to supply the requirements for both household and garden purposes, if the diameter of the wells and the size of the water-tanks are made to provide against the contingency of an occasional period of calm. Experience has demonstrated that a well of ten feet diameter, with a good wind-mill and pump, and a tank of 12,000 gallons capacity, will, with judicious management, afford water enough for an acre of lawn, besides what is needed for domestic purposes. As a consequence, wind-mills are quite a feature of Oakland.

The quality of ordinary well-water is not uniform. Some of it is hard, but, with rare exceptions, it is all pleasant to drink. Judging from the uniformity of the substratum of indurated sand and clay which underlies the site of Oakland, we are inclined to believe that soft water can be obtained in all parts of the city, if wells are sunk to the proper depth, and the curbing cemented so as to keep out surface water.

The stratum of indurated sand and clay, above mentioned, is impenetrable to surface water, and makes an admirable filter for water percolating through it. Hence, if the curbing of wells be cemented to a proper depth, and packed with clay on the outside, on a level with the "hard pan," even the proximity of cess-pools cannot impair the purity of wells.

All efforts, in Oakland, to obtain overflowing artesian wells, have failed, but they have resulted in the next best thing, to wit: inexhaustible wells of soft, pure water which comes within a few feet of the surface. We know of four such wells in as many different parts of the city.

The result of experimentation in artesian well-boring indicates the existence of a stratum of pebbles and red gravel, at a depth of less than one hundred feet, through which water percolates freely, under a sufficient pressure to bring it *near* the surface; and it is money thrown away to sink an artesian well below the stratum of gravel. The water obtained from the latter source is soft and pure.

THE CONTRA COSTA WATER COMPANY

Furnishes the following statement respecting the water now being supplied from the mountain range back of Oakland:

"The water is collected at a point five miles from the city, near the head of Temescal Creek, where two streams flow constantly into a reservoir. The watershed supplying the streams, above the reservoir, embraces an area of three thousand acres, too precipitous for cultivation. It is estimated that a rain-fall of twelve inches upon this water-shed will furnish more than one thousand millions of gallons. The reservoir capacity is now small, but is being increased to about two hundred millions of gallons, and can be further increased as occasion requires."

The energy exhibited by the Company is highly commendable. It has already laid about thirty miles of pipe, ranging in size from three to fourteen inches. The estimate of the water supply obtainable from this source, is three millions of gallons per day.

The drought of the present year (1871) has demonstrated the uncertainty of the Company's calculations; and it has been obliged to resort to artesian wells, and steam power, to furnish its patrons with water. The charges, for domestic purposes, are the same as at San Francisco.

WATER RESOURCES.

In Amador Valley, thirty miles from Oakland, there is an abundance of soft, pure water, sufficient to supply a population exceeding half a million. The water-basin is the receptacle of six hundred square miles of adjacent country, with its tributary streams.

The water exists in a Tule Lake, partly subterranean, five hundred feet above tide level, surrounded by hundreds of natural wells, which are full to the brim in the driest seasons. During ordinary wet seasons, these wells overflow and inundate a large surface. The sources that supply the lake are constant — the most important of which are the Los Positas, in the Livermore plain; the Arroyo Mocho, and the Arroyo del Valle, on the east and south; the Arroyo el Alamo, Arroyo de la Tasajera, the Los Alamos, and San Cayetao from the north. Most of these are living streams flowing into the lake. There is but one outlet to this water —at the south-west end of the lake—debouching from which, the water forms the Laguna Creek that flows southerly, parallel with the Central Pacific Railroad, six miles to Suñol Valley. There, it forms a junction with the Alameda Creek. The water from the two sources forms a large and beautiful stream which meanders, side by side with the railroad, through the Alameda Cañon to Vallejo's Mill. (See map.) Thence, it flows south-westerly, by the town of Alvarado, to the Bay of San Francisco.

By diverting the water at the junction of the streams, and conveying it along the mountain-sides, through the cañon, five miles to Vallejo's Mill; thence, westerly, along the foot-hills to Hayward's; the San Lorenzo Creek, a large and rapid stream, could be made tributary. Four miles nearer Oakland, is the San Leandro Creek, likewise available as a tributary, and which, alone, would furnish a supply of water for a population of fifty thousand.

The water from these sources would not only afford Oakland an ample supply, for many generations, but the places on and near the line of approach, including Niles' Station, Decoto, Alvarado, Hayward's, San Leandro, Alameda, and Brooklyn, could reap a similar benefit.

The foot-hills present the convenience for conveying the water from the above-mentioned sources to a grand reservoir back of Oakland, one hundred feet above the level of the highest part of the city.

North-west of the city, there are also sources whence supplies are obtainable, the most important of which are the San Pablo Creek, fifteen miles distant, and the Wildcat Creek, near the State University grounds. The water from both could be brought to the grand reservoir.

We are not prepared with estimates of the cost of obtaining this great water supply; but from information given us by skillful engineers who have examined the ground, we can safely say that it would be trifling, in comparison with its importance.

The subject is already attracting the attention of enterprising men, and is worthy that of our city authorities.

STREETS.

The aggregate length of all the streets in Oakland, is, in round numbers, one hundred and five miles, of which fourteen miles have been macadamized and otherwise improved. The streets are generally eighty feet wide, and in most cases cross each other at right angles. Broadway, the principal thoroughfare, is one hundred and ten feet in width, the sidewalks being twenty feet wide. The streets are macadamized with a hard, blue trap rock, of a very superior quality, which is found in great abundance in the immediate vicinity of the city.

The following are streets, and portions of streets, that were graded, macadamized, and curbed, during 1870:

Oak Street, from Seventh to Twelfth.....................	1,320 feet.
Julia Street, from Eighth to Ninth.....................	200 "
Alice Street, from Eighth to Fourteenth.....................	1,520 "
Washington Street, from Eighth to Fourteenth.....................	1,550 "
Clay Street, from Eighth to Tenth.....................	480 "
Brush Street, from First to Twelfth.....................	2,760 "
Market Street, from Seventh to Forty-second.....................	4,420 "
Sixth Street, from Castro to Franklin.....................	2,040 "
Seventh Street, from Broadway to Franklin.....................	300 "
Ninth Street, from Clay to Oak.....................	3,020 "
Tenth Street, from Broadway to Alice.....................	1,360 "
Fourteenth Street, from Broadway to Washington.....................	300 "
Total.....................	19,240 feet.

The average cost of macadamizing is estimated at 6¼ cents per square foot. Nineteen thousand two hundred and forty lineal feet of roadway and crossings, converted into square feet, gives:

926,840 feet, at 6¼ cents.....................	$57,927 50
35,199 feet curbing, at 12½ cents.....................	4,233 88
Engineering, advertising, and culverts.....................	3,000 00
Total cost.....................	$65,161 38

GRADES.

The city of Oakland is situated on a peninsula extending about one and one-half miles from north to south, and two and one-half miles from east to west. It is bounded on the south and east by San Antonio Creek, on the west by the Bay of San Francisco, and on the north by the charter line, established by Act of the Legislature, in May, 1852. The highest ground in the city is found about midway between the northerly and southerly boundaries, and is thirty-eight feet above the level of high tide. From this water-shed the ground slopes with remarkable uniformity, southerly and easterly, to the estuary, and northerly, to a depression near the charter line, and to the salt marsh along the shore of the bay. Sufficient fall is everywhere obtained for surface drainage, and no serious difficulty is encountered in establishing surface grades.

Something over a year ago, the Common Council appointed a Board of Engineers, "to examine the plans and profiles of the city of Oakland, to suggest changes, if any they may deem necessary, and to report a plan of street grades, lines, and a system of sewerage for the whole city." The Board was composed of George F. Allardt, Chief Engineer of State Tide Lands; Prof. George Davidson, Assistant U. S. Coast Survey; George E. Gray, Consulting Engineer Central Pacific Railroad Company; Milo Hoadley, President of the late Board of Engineers of San Francisco, and William F. Boardman, late City Engineer of Oakland.

It has seldom been the fortune of any city to obtain the combined services of the same number of engineers so eminent in their profession and so well qualified in every respect to deal with the important problems submitted to this Board.

In due time they presented an elaborate report, and all street improvements and other public works are now executed in accordance with their recommendations. On the uplands, the grades adopted by the Board conform to the natural surface of the ground, so far as is consistent with an efficient system of drainage and sewage. On the salt marshes and tide lands along the water-front, while due regard is given to the future commercial requirements of the city, the grade is not placed so high as to be onerous or oppressive to the property-owners.

SEWERS.

It is proposed to construct two main sewers of sufficient capacity to receive the surface and sewer drainage of the entire peninsula. One, along or near San Antonio Estuary, and the other through the depression near the charter line on the north. The aggregate length of the two sewers will be about five miles. The tidal waters retained in Lake Peralta, at the eastern terminus of San Antonio Creek, will be used for the purpose of flushing the main sewers at stated intervals. The bottom of the upper end, or inlet, of either sewer will be placed one foot below high tide; the bottom of the outlet at the Bay, one foot below low tide — giving a fall of six and a half feet, which is sufficient to keep the sewers free from all deposits.

Surface water, and house sewage, will be conveyed to the main sewer by means of smaller lateral sewers of cement pipes, twelve inches in diameter. Gradients of one in one hundred and fifty can be obtained in the most unfavorable localities. The projected system of sewage is admirable, and its cost will be unusually small.

STONE QUARRIES.

There are inexhaustible supplies of basaltic trap rock found in the foothills, within a distance of from two to three miles north-easterly from Oakland. There are now two macadamizing companies engaged in paving the streets of Oakland and Brooklyn, with rock obtained from the above-mentioned source, and they employ about one hundred men. Both companies have machines for crushing the material and graduating its size. The crushing capacity of each is from seven to ten tons per hour. The character of our paving far excels the old fashioned macadamizing, and the quality of the material, for paving purposes, is not surpassed elsewhere in the world. The cost of paving is mentioned on another page.

Ledges of excellent sandstone are also found in the hills, at a short distance beyond where the material for macadamizing is obtained, and the stone is being used for building purposes.

RAIN TABLE FOR OAKLAND, SHOWING THE RAIN-FALL EACH MONTH FOR TWENTY YEARS.

MONTHS.	1850.	1851.	1852.	1853.	1854.	1855.	1856.	1857.	1858.	1859.	1860.	1861.	1862.	1863.	1864.	1865.	1866.	1867.	1868.	1869.	1870.
August		0.1							0.1				0.1		0.2			0.1			
September		1.0					0.1									0.2	0.6	0.2	0.1	0.1	
October	0.2	0.2	0.8	0.1	2.1	2.1	0.5	0.9	3.4		0.9	3.8	1.4		0.1	0.1	3.1	0.6	0.2	1.3	1.3
November	1.3	2.2	5.3	1.4	0.4	1.2	2.9	3.0	0.5	5.4	0.2	6.1	1.1	2.5	7.6	3.1	2.7	3.1	1.2	1.1	0.3
December	1.1	7.1	11.9	2.1	0.4	5.4	4.0	4.2	4.8	1.5	4.8	6.1	2.7	1.7	6.9	0.6	13.1	12.1	4.3	4.2	2.9

MONTHS.	1851.	1852.	1853.	1854.	1855.	1856.	1857.	1858.	1859.	1860.	1861.	1862.	1863.	1864.	1865.	1866.	1867.	1868.	1869.	1870.	1871.
January	0.6	4.1	4.3	4.5	8.4	2.1	4.4	1.0	1.1	1.2	18.1	3.3	1.3	3.9	11.0	6.6	9.6	6.4	0.7	3.1	3.1
February	0.4	0.1	8.4	4.6	4.6	8.6	1.3	5.2	0.3	2.8	6.1	3.3	3.3	0.8	1.5	6.2	6.2	4.0	0.6	3.8	3.8
March	1.9	6.4	3.2	4.3	4.0	1.6	3.9	3.1	2.5	3.4	1.1	2.4	1.7	1.4	0.6	2.6	6.2	6.4	2.2	4.1	1.3
April	1.1	0.2	5.1	3.3	5.6	3.2	1.1	0.3	1.0	2.6	0.3	0.9	2.9	1.1	0.7	0.1	1.1	2.2	2.2	0.1	1.9
May		0.3	0.1	0.1	2.2	0.9	0.1	2.0		0.7	0.7	0.4	0.4	0.5	0.4	1.8	0.1		0.2	0.1	0.2
June	0.7			0.1					0.3		0.2	0.2				0.2		0.2		0.4	
July																					
Total	7.1	18.2	33.5	23.0	24.1	21.2	20.0	19.0	19.8	17.1	14.6	38.0	15.2	8.5	21.3	21.2	32.2	40.5	21.6	15.3	13.5

THE PHILOSOPHY OF GRAIN GROWING.—We copy the following from the *San Francisco Evening Bulletin:* "The philosophy of grain growing in California is as follows: If the early rains supply moisture enough to enable the farmers to seed the ground, that is amply sufficient. Enough is better than a feast in this respect, for the farmer can seed more acres. The success of the crop depends on the spring rains; and if the rain-fall in the spring is sufficient to cause the moisture from above to meet the moisture below, the crop is assured. The moisture rising by capillary attraction is the security; whereas, a dry streak between the moisture above and below is fatal. The fall of rain in November and December, 1850, was but 2.40, which is a full inch less than the rain-fall during the corresponding months of 1870. January, 1851, yielded only 0.58; February, 0.12; but March yielded 6.40; April, 0.19; and May, 0.30. The spring rains did not commence in earnest until March; yet, what was the result? Mr. Beard, of San José Mission, invested $10,000 that season in planting wheat, and in February he would have been glad, as he says, to have sold out for a song. The March rains were sufficient to carry the moisture down to meet the moisture below, and Mr. Beard reaped a larger crop that season than in any following year." Except in the Coast counties, the spring rain-fall of 1871, in our grain districts, was insufficient to obliterate the "dry streak:" hence, the failure of crops. Having experienced three comparatively dry seasons, successively, the "oldest inhabitants" predict a plentiful rain-fall during the approaching winter.

SANITARY AND MORTUARY.

From a sanitary point of view, Oakland stands unrivaled among the cities of the Pacific slope. This is a bold assertion; nevertheless, it is confirmed by official records.

We shall not enumerate the causes which render Oakland so eminently desirable as a place for family residences, but we shall proceed to prove that not another of the principal cities in the State can claim such exemption from sickness and death.

We quote the recent census reports respecting the population of the several cities; and the mortuary statistics are summarized from the reports of Dr. Logan, President of the State Board of Health, published in the *San Francisco Medical Journal.*

NUMBER OF DEATHS DURING THE YEAR ENDING JUNE 30, 1871.

CITIES.	1870.						1871.						Total......	Population.
	July...	Aug...	Sept...	Oct...	Nov...	Dec...	Jan...	Feb...	March...	April...	May...	June...		
San Francisco	298	281	264	309	347	266	298	245	227	232	226	221	3,214	150,361
Sacramento	31	31	29	50	46	33	28	24	31	24	39	26	392	16,298
OAKLAND	7	10	7	13	9	12	12	13	7	9	10	8	117	11,104
Stockton................	23	14	16	17	16	18	22	9	9	4	12	20	180	10,033
San José................	16	14	21	16	10	24	19	18	9	*9	*9	14	179	9,091

* The deaths at San José during April and May, 1871, do not appear in Dr. Logan's Reports; and to avoid injustice, as between Oakland and San José, we have inserted figures corresponding with the minimum reports of other months.

Discarding the decimals, the above exhibit shows, during the twelve months, one death in San Francisco to about every 46 inhabitants; in Sacramento, one to 42; in Oakland, one to 95; in Stockton, one to 56; and in San José, one to 51.

It is but fair to deduct from the deaths set down to San Francisco, the number which resulted from suicides and casualties; and it should be borne in mind that many persons afflicted with disease contracted elsewhere, visit San Francisco for medical treatment; and the proportion of these who die, should also be deducted from her mortuary reports, when we are comparing sanitary conditions. Deducting 143 deaths from her 12 months' report to cover the former, and 12 *per cent.* from the remaining 2,702, to compensate the latter, the result will show nearly *double the number of deaths in San Francisco, in proportion to the population, as have occurred in Oakland.*

The comparison between Oakland and the other cities, is no less wonderful; and, considering that Oakland is a favorite resort for persons suffering from disease, the above exhibit will astonish the people of Oakland little less than persons abroad.

DURATION OF SICKNESS.

Before concluding, we will refer to a collateral fact, alike unprecedented in sanitary annals, yet, supported by incontestable evidence. For the purpose of getting information concerning the average duration of sickness in Oakland and vicinity, we have examined, by permission, the books of two of our most prominent physicians. We took the aggregate of the visits made by the two physicians for six months, and divided the sum by the total number of patients visited. The result was an average of *four and one-third* visits to each case. By leaving out of the calculation several desperate cases, the average would be considerably less. The books of the aforesaid physicians will be cheerfully submitted to the inspection of any respectable practitioner who may think we have committed an error.

Is there another city in the United States whose population enjoy such exemption from sickness and death? If there be one, sign-boards should be erected on every highway and lane approaching it, warning physicians and undertakers of the danger from starvation attending a residence within.

At the request of the Publication Committee, we have investigated the data of the foregoing article, and found it to be correct. CLINTON CUSHING, M.D., Pres't. Alameda Co. Medical Association.
JOHN C. VAN WYCK, M.D., Librarian.
OAKLAND, Oct. 1, 1871.

DRIVES AND SCENERY.

There are few places upon earth which are more inviting to those fond of outdoor exercise, than Oakland and its vicinity. If it be true—as it unquestionably is—that the Bay of San Francisco is the finest and most picturesque in the world, not even excepting the Bay of Naples, and the magnificent harbor of Rio Janeiro, it is no less true that the site of Oakland affords the most beautiful view of that Bay, and the most delightful of the valleys by which it is environed. Here, the Coast Range, generally so abrupt and rocky, recedes gradually into a vale miles in width, and slopes with a gentle declivity to the waters of the Bay that bathe its borders with the health-inspiring ripples of the Ocean, just visible through the opening of the Golden Gate. Eastward, the summit of Mount Diablo presents one of the loftiest peaks from San Diego to Shasta Butte. Westward, gleams the broad bosom of the Bay, bordered in the distance by the triple hills of San Francisco, the blue summits of the San Bruno Range, and the slumbering valleys of San Mateo. Northward, stretch the fruitful orchards of San Pablo, the green hills of Carquinez, and the fairy islets of Golden Rock and The Sisters; while southward, the old Mission of San José looms up in the distance like a glimpse of Eden; and the most fertile of hills, and dales, and plains, commingle in the view, assuring the spectator that no land on the globe unites in itself blessings more varied, or landscapes more enchanting, than those which greet the eye from the flower-enameled plain of Alameda.

Here, are no toll-roads, to check adventure and tax the pleasure-seeker with their oppressive exactions. There are no craggy precipices to climb, or soft morasses to cross; but the country is intersected with highways attesting the genius of MacAdam, and leveled like the thoroughfares of Holland. Are you weary of city life, and require the mountain air to invigorate your frame? Scale the summit of Mount Diablo! Are you ill, and need the waters of old Ponce de Leon to reanimate you with the vigor of perpetual youth? Go and bathe in the fountains of the old Mission San José! Are you fond of sport? Shoulder your gun and gather quail from the foothills, or rig your fishing-tackle and bait for smelt or silver-fins, for trout or perch, off the ends of our piers, or in the shady nooks of the San Leandro! Are you a lover of Nature? Mount your horse, and thread the grounds of the State University! Visit the gems of the foot-hill farms! Climb the gentle acclivities of the Coast Range! And, turning suddenly in the saddle, cast your eyes on the slumbering landscape at your feet! Where upon the broad earth can your gaze meet with so enchanting a spectacle? Vineyard, orchard, and garden; fountain, bay, and ocean; plain, meadow, and mountain, blend in a unison so perfect, that you feel there can be no spot where Nature presents greater inducements for homes, than the gorgeous queen of the valleys, the beautiful bride of the Bay, the flourishing city of Oakland.

WHAT NATURE HAS DONE.

She has given us a climate unsurpassed in the world—preserving the health of those who are not afflicted, and imparting health to those who are.

She has given us a soil, in harmony with the climate, which affords sustenance to nearly every description of plants and trees.

She has given us a site for a city which, comparatively speaking, is already graded; she has ornamented it with a profusion of majestic oaks, and sent larks and linnets to perch upon the boughs and delight us with their warbling.

She has given us a never-failing supply of pure water within a few feet of the surface, and she guards it from contamination by a formation of sand and clay, impervious to surface water.

She has placed, within a convenient distance, inexhaustible supplies of pure water which may be conducted, by gravitation, alone, to the tops of our highest buildings.

She has placed, close at hand, ledges of stone admirably adapted to building and macadamizing.

She has surrounded us with scenery which delights the eye, expands the mind, and animates the spirits.

She has given us, in common with San Francisco, one of the finest harbors in the world; and she has banished the *teredo* from our shores.

She has given us a solid foundation for buildings and wharves, from high-water mark to ship-channel; and she deposits her mud elsewhere.

She has made depressions in the mountain ranges which lead the locomotive to our wharves to meet the commerce of the ocean; and has ordained Oakland as the great terminus of the railroad system of the Pacific Coast.

STREET RAILROADS.

The contour of Oakland and the surrounding country, being almost level, or gently undulating, is peculiarly well adapted to horse-railroad enterprises. There is one already in successful operation, extending from the foot of Broadway to Telegraph Avenue, and thence to Temescal Bridge. Its franchise extends to the State University grounds. Its present track is three miles in length, and the cars and horses used by the road company compare favorably with those used in San Francisco. The success of the enterprise has stimulated the projection of other horse-railroads, among the most important of which is one designed to connect Fruit Vale and Brooklyn with the University grounds, and one to connect the San Francisco and Oakland Road with the University grounds, via Peralta street. The latter will be built and owned by the C. P. R. R. Company. The Oakland and San Pablo Avenue Company, and the San Pablo, Webster Street, and Alameda Company, have also located routes of great importance; and the roads already projected will form, when completed, a cordon of iron rails which will afford the people of Oakland, and the neighboring towns, cheap and constant facilities of communication with each other, and with the State University.

OAKLAND GAS-LIGHT COMPANY.

This Company has fourteen miles of "mains" already laid in Oakland, besides extensions to and about the town of Brooklyn. The present capacity of the works is one hundred thousand feet per day, and the quality of the gas, is not surpassed by that of any other company in California. There are few, if any, cities in the United States of an equal number of inhabitants, wherein such an extent of gas-mains has been laid. The quantity of gas consumed is not commensurate with the extent of the mains; but that militates against the Company, and in favor of property-owners, and of those who desire to build houses and to enjoy the luxury of gas-light.

A Pneumatic Gas Company has obtained a franchise for laying pipe in Oakland; but whether or not its pipe will be lighted, remains to be seen.

LAND TITLES.

The stability of the title to real estate in Oakland and Brooklyn townships, recommends it strongly for investment and homestead purposes. It is a fundamental principle in English and Spanish law, derived from the maxims of the feudal tenures, that the King was the original proprietor of all land in the Kingdom, and of all territories acquired (like California) by discovery and colonization, and that he was the only and true source of title. In the United States, the same principle has been adopted. All valid individual titles to land in California are, therefore, derived from the Government of the United States, and the State of California —from the latter subordinately, and only for land covered by tide-water; or, from the Spanish Crown, prior to the 28th of September, 1821—the day recognized in law as the date of the independence of the Mexican nation; or, from the Government of Mexico up to the 7th of July, 1846, when the United States took possession of this State which was subsequently ceded to them by the Treaty of Guadalupe Hidalgo, February 2d, 1848—by which treaty all governmental grants, previously made, were confirmed.

Thus, was the title to the lands in the city of Oakland, and the town of Brooklyn, together with that of the surrounding country, comprising about twenty-five thousand acres, derived from the Mexican Government, through a grant made in 1820 to Don Luis Peralta, in recognition of his meritorious services in the conquest of California.

Peralta divided his rancho, first, by actual partition in 1846, and afterward (in 1851) by will, between his four sons, José Domingo, Vicente, Antonio, and Ignacio, whose titles have since been recognized and confirmed by the United States Courts. Efforts were made to assail and cloud fractional parts of the title of these brothers, but the Courts have rejected, and declared invalid, all adverse claims.

No real estate can be held under a better title than that which is derived from the brothers Peralta.

THE PRICE OF HOMESTEAD SITES.

In all places where people most do congregate, the active competition for the possession of land, causes the value of real estate to rise with the increase of inhabitants. Thus, has property in San Francisco become very valuable, mostly in the eastern portion of the city, specially devoted to business in its various branches; thence, southerly, over flat lands; and westerly, over hills and through dales, in all inhabitable directions, where year by year dwellings multiply.

But this increase in value is not confined to the limits of the metropolis. It spreads for miles over neighboring localities which are attractive for family residences, as they are brought nearer by means of increasing facilities for travel.

It is so with the surroundings of New York, and all large cities; and the history of the last few years plainly indicates that the same causes are producing like results here. The attention of those whose interests or preferences have called them to San Francisco, has, of late years, been more and more directed, for climatic and other reasons, toward suburban retreats, chiefly in the direction of Oakland and vicinity. Values have consequently increased, but, apparently, not in proportion to the progress in population and improvements, nor to the prospective importance of the locality.

The object of this article is to invite attention to the very considerable difference which still exists in the value of residence property in San Francisco, as compared with that in Oakland and Brooklyn. Various considerations may lead people to prefer a residence outside of the great city to one within, and not the least among these is the larger quantity of ground obtainable for the same amount of money.

For this purpose it will be useful to compare the value of residence property in the places named, for lots of different depths, on streets of different widths—items which enter largely into calculations of value.

It is evident that no very precise comparison of one locality with another can be made, as no two localities can be said to offer exactly the same advantages; nor, owing to the diversity of individual appreciation, are they susceptible of being judged by the same standard.

The information, herewith submitted, has been obtained from reliable sources. Opinions on values will always differ, more or less, but the valuations have been carefully made, though necessarily in a general way, and are intended to represent prices which can be realized when opportunities for sales occur. All quotations are stated per foot frontage for inside lots—corner lots being worth from ten to thirty per cent. more.

In San Francisco, on streets 82½ feet wide, like Mission, Howard, and Folsom Streets, property ranges, for lots 80 to 90 feet deep, from Fourth to Seventh, at $125 to $200 per foot frontage; and lots beyond Seventh, to Fourteenth, at $75 to $100; farther southerly, to Twentieth Street, $60 to $75, and on Valencia, $80 to

$90; on Van Ness Avenue, $120 to $150; on the other streets, in the Hayes and Beideman tracts, about 69 feet wide, lots 120 feet deep are worth $60 to $100 per front foot.

In Oakland, east from Market Street, lots 100 feet deep on all the 80 feet streets north of Railroad Avenue or Seventh Street, sell for $27.50 to $50 per foot frontage; and south of Seventh Street, at $22.50 to $30. On Adeline and Market Streets, both 80 feet wide, lots 125 and 130 feet deep, between Seventh and Twenty-second Streets, bring $27.50 to $45 per front foot.

Again, in San Francisco, on Stevenson, Jessie, Minna, Natoma, and similar streets, only 35 feet wide, lots 70 to 80 feet deep, between Fourth and Seventh Streets, bring readily $50 to $60 per foot frontage, and from Seventh to Tenth, about $40.

Oakland and Brooklyn have no streets of such limited width—the narrowest measuring 60 feet. On the 60 feet streets in Oakland, property sells as follows: North of Seventh, to Fourteenth, between Market and Adeline, $30 per foot frontage, 125 feet deep; from Fourteenth to Eighteenth, between Market and Adeline, 125 feet deep, $16 to $22.50; between Kirkham and Peralta, north of Fourteenth Street, 104 feet deep, $12 to $20; between Peralta, Pine, Eighth, and Twelfth Streets, near the Point, lots 135 feet deep, $22.50 to $25; between Adeline and Peralta, Seventh and Fourteenth, lots 125 feet deep, $20 to $22.50; at the Point, both north and south of Seventh Street, lots 100 feet deep, $22.50 to $30; north of Twenty-second Street and west of the San Pablo Road, lots 125 feet deep, $10; east of the said road, lots 110 feet deep, $15 to $20 per foot frontage.

In Brooklyn, property on 60 feet streets is worth: West of Walker, and south of Humbert Streets, lots 150 feet deep, $10 to $15 per foot frontage; north of Hepburn Street, lots 140 to 150 feet deep, $5 to $10.

The reader will bear in mind that reference has been made solely to residence property, and our allusions to San Francisco values do not refer to certain favored localities where even residence property is held as high as $300 per front foot. Respecting business property, those who desire to purchase, may seek information for themselves. It is hardly necessary to say that business property is far more valuable in San Francisco than in Oakland.

BUILDING IMPROVEMENTS IN OAKLAND.

On January 2d, the *Oakland Daily Transcript* published a table showing the location and value of the buildings erected in this city during the year 1870, from which it appears that 615 houses were built, at a total cost of $1,405,150. Since the first of January, 1871, a very large number of buildings have been commenced, and the improvements for 1871 are very likely to exceed in value those made in 1870, by at least half a million dollars.

COST OF BUILDING.

The cost of building in Oakland is somewhat less than in San Francisco. The lumber-yards, and the planing-mills, are conveniently located, and the ground which they occupy is much less valuable than that occupied by similar establishments in San Francisco. Bricks and stone are obtainable cheaper here than at San Francisco; castings are supplied by the local foundry; and, generally speaking, no grading or filling is required.

MANUFACTURING PROSPECTS.

The map of Oakland shows the outlines of the contemplated improvements of the Water Front Company. The most important features of the project are the dredging of the Bar at the mouth of the San Antonio Estuary, the cribbing of both banks, from ship-channel to the head of the southerly arm, a distance of over five miles, and the widening and deepening of the channel where necessary.

There will be a continuous wharf between the water and the first tier of blocks on the north bank of the channel, from its mouth to Broadway Street. A wide street is provided for, in the rear of the tier of blocks, to accommodate as many rail tracks as may be needed. These tracks will lead to the main trunk of the C. P. R. R. Thus, a manufacturing establishment situated upon any of the aforesaid blocks will be able to receive or deliver freight at "ship's tackles," at the front doors, and to load or unload cars at the back doors. If desirable, "turn-outs" can be laid from the street, passing through the building to the water; and it requires no gift of prophecy to predict that, as the projected improvements are made, the heavy manufacturing business of the Bay counties will concentrate where such facilities for economizing are provided: and there is not another place about the Bay where it is possible to provide them. The perusal of our remarks under the head of "The Estuary of San Antonio," will give the reader additional light concerning the vast prospective importance of the manufacturing interests of Oakland.

BRIDGING THE BAY.

Some of our San Francisco neighbors seem much alarmed about commercial prospects at Oakland, and are indulging extraordinary vagaries respecting things which they deem necessary to save their city from decay.

The fact is, San Francisco is more interested than Oakland, in commerce at Oakland. That is to say, 150,000 people are more interested than 11,000 people, in reducing the cost of handling exports and imports. For example, unless we can compete with other countries, in shipping grain to distant markets, the cultivation of grain in California, except for home consumption, will cease, and every branch of industry and trade in San Francisco would suffer. On the contrary, if, by means of machinery, and the economical handling of the grain crops, farmers have the assurance of realizing a profit, they will seed more land, and every branch of industry and trade in San Francisco will be stimulated by the success of the farmers.

This proposition is as simple as "rolling off a log;" yet, a portion of the press and of the people of San Francisco are exercised at the economical arrangements at Oakland, for handling our export products; and are proposing to tax the community for the purpose of supplying other, and far less economical, arrangements elsewhere. They are even advocating the vandalism of destroying half the value of a great harbor which belongs to the Commerce of the World, in the vain hope of forcing business into an unnatural channel.

They are horrified at the laying of a "gas-main" across Mission Creek, where hogs wallow at low tide; but are in ecstasies at the thought of cutting off more than *ninety-two square miles* of the navigable waters of the Bay of San Francisco, from free commerce with the ocean, by constructing a bridge from Alameda, or Oakland, to San Francisco.

Nor is this all: The conductors of the San Francisco press are well aware that solemn warnings have been uttered by the highest hydrographical authorities in the United States, against obstructing the currents of the Bay, in any way that might decrease, to a great extent, its tidal area; for, upon the tidal area, depends the volume and scouring effect of the tidal flow over the Bar, at the entrance of the harbor, and the depth of water upon it.

In view of this warning, and considering that it is impossible for engineering skill to predetermine the effect of placing fifty, or more, immense piers, in a line across the channel of the Bay, it seems extraordinary, to say the least, that respectable journals in San Francisco should advocate such a project.

San Francisco cannot afford the experiment. New York and Boston cannot afford it. The merchant marine of California, and the farmers of California, protest against it.

We shall now proceed to enlighten the reader respecting the pecuniary benefits, and the commercial advantages, which San Francisco might reasonably expect from the construction of a bridge; and we challenge any engineer to discover a material error in the following estimates, by Geo. F. Allardt, C. E., who furnished them by request. Mr. Allardt is recognized by Engineers as one of the foremost men in the Profession:

Estimated Cost of Bridging the Bay from San Francisco to the Alameda Shore—Distance, five miles (26,400 feet), of which three miles (15,840 feet) will extend across ship-channel; (from 18 to 60 feet in depth at low-tide); and two miles (10,560 feet) across shoal water on the Alameda shore.

First, a wooden bridge throughout: two miles of pile trestling in shoal water, and three miles of Howe truss in deep water, supported on *pile-piers*, with spans of 200 feet each, including three turn-table spans, or "draws." Bottom of trusses to be ten feet above high-water, in the clear.

10,560 lineal feet of pile-trestling, @ $20	$211,200
79 pile-piers in deep water, @ $4,000	316,000
15,840 lineal feet of Howe truss, @ $60	950,400
Extra expense on three turn-table spans	25,000
	$1,502,600
Add 10 per cent. for superintendence and contingencies	150,260
Total cost	$1,652,860

Or $62.63 per lineal foot.

Second, the same, except with *stone-piers*, across the deep water, in place of *pile-piers*.

10,560 lineal feet of pile-trestling, @ $20'.................... $211,200
15,840 lineal feet of Howe truss, @ $60...................... 950,400
80,400 cubic yards of masonry in 79 piers, @ $40............. 3,216,000
Extra expense on turn-table spans and piers................. 50,000

                                                          $4,427,600
Add 10 per cent. for superintendence and contingencies......... $442,760

Total cost.. $4,870,360
Or $184.48 per lineal foot.

Third, pile-trestling for two miles, as before ; stone-piers for three miles, across deep water, and iron trusses, in place of the Howe truss.   Spans 200 feet.

10,560 lineal feet of pile-trestling, @ $20 .... $211,200
15,840 lineal feet of iron truss, @ $200 3,168,000
80,400 cubic yards of masonry in 79 piers, @ $40............. 3,216,000
Extra expense on turn-table spans and piers.................. 50,000

                                                          $6,645,200
Add 10 per cent. for superintendence and contingencies 664,520

Total cost... $7,309,720
Or $276.88 per lineal foot.

Fourth, a first-class high bridge, with stone-piers and iron superstructure throughout, placed 100 feet, in the clear, above high-tide in ship-channel, and with ascending gradients of 50 feet to the mile, across the shoal water on the Alameda shore, and in Mission Bay at San Francisco.   Spans 300 feet each.

160,900 cubic yards of masonry in 53 piers in deep water (3 miles)
    @ $30..........................'................... $4,827,000
32,600 cubic yards in 35 piers on the gradient on the Alameda
    shore (2 miles) @ $30............................ 978,000
32,600 cubic yards in 35 piers on the gradient on the San Fran-
    cisco shore, @ $30.............................. 978,000
36,960 lineal feet (7 miles) of iron superstructure for double track,
    wagon-road, and foot-passengers, @ $225........... 8,316,000

                                                         $15,099,000
Add 10 per cent. for superintendence and contingencies 1,509,900

Total cost...$16,608,900
Or $449.38 per lineal foot.

For the purpose of comparison, we quote, below, the cost of several long bridges, the average of which is over $750 per lineal foot:

NAME.	LOCALITY.	Height above high-water, feet.	Length, feet.	Cost.	Cost per lineal foot.
Britannia	Menai Straits	102	1,841	$3,009,325	$1,635
Niagara (suspension).....	Niagara Falls...........	245	1,290	400,000	310
St. Charles..............	Missouri River..........	80	6,570	1,815,000	276
East River..............	New York to Brooklyn...	103	5,625	7,000,000*	1,244
Louisville...............	Ohio River.............	56	5,280	1,600,000	303

* Estimated.

Even if it were permissible to place a line of piers across ship-channel in the Bay, nothing less than a permanent first-class structure, one hundred feet above high tide, over ship-channel, would be allowed; and the gradients of the approaches should not exceed fifty feet to the mile, if designed for freight trains. The distance from the Alameda shore to ship-channel (two miles) would accommodate the eastern approach; but, if the western approach started on the same level as the eastern, the westerly end of the bridge would be *two miles* distant from the San Francisco water front. To compensate for the height of the bridge, five miles must be added to its length, to make it equivalent to a level way, for railroad purposes. (See note, page 3, "Equating for Grades"). If the reader will figure the result, he will find that, for all practical purposes in railroading, the distance from Alameda, or Oakland point, via such a bridge, to the water front of San Francisco, would be *fourteen miles.* Thus, San Francisco, by the expenditure of over sixteen millions, could double the distance and time between herself and Oakland.

But it is not designed to build a first-class, *high bridge*—it would not pay. Indeed, our introductory remarks are based upon the assumption that a *low bridge* is contemplated. We refer the reader to the foregoing estimates, if he desires to fix in his imagination the cost of the undertaking, and we shall proceed to estimate the benefits that might accrue from it, leaving out of the question the damages that would result from destroying free navigation in the Bay, and the peril of shoaling the Bar at the entrance of the harbor:

First—Building lots, and homestead sites, in Alameda County, would increase largely in value, in anticipation of an exodus of families from San Francisco who object to steamboat travel.

Second—Considering that the freight and passenger trains of the Central, and the Southern Pacific, will soon approach Oakland from the north; if the bridge started from Oakland Point, passengers and freight might be carried thence to San Francisco, by rail, five minutes quicker than by rail and boat—providing no "draws" were open, as frequently there would be. But Oakland would never consent to obstructing the Estuary of San Antonio from free commerce with the ocean. Hence, the easterly end of the bridge, if constructed, would be at Alameda, or at a point farther south.

Third—Assuming that it would be at Alameda point: the distance from Oakland point (where the Oakland and Banta Branch will terminate) to Alameda point, via the most available crossing of the Estuary, is five miles. It follows that passengers might reach the San Francisco shore at Mission Bay—supposing the bridge should terminate there, and no "open draws" were encountered—*almost* as soon as they could reach the hotels by the Oakland Ferry. But the existence of the bridge would not control, in the slightest degree, the movement of our export products. The great freight route must intersect the Oakland wharf, as the map shows; and it is as reasonable to expect that grain, for export, would be transported from San Francisco to Oakland, via the proposed bridge, as to expect it to be moved in the opposite direction. Mr. Friedlander, and San Francisco exporters generally, having grain arriving at the Oakland wharf, would decline to incur the needless risk and expense of transporting it from the Oakland docks to the San Francisco docks, to gratify a sentiment. It may be added that the proposed bridge could never be used for passenger travel, except for that between San Francisco and Alameda counties. The liability of detention by "open draws" would render

3

it impossible for the railroad companies to run "through trains," by that route, "on time."

If the hotels, churches, stores, and theatres of San Francisco, were removed south of the bridge, it would compensate, in a measure, for some delays; but, in that event, it would become necessary to remove the Golden Gate to a point south of the bridge—otherwise, the rapid increase of commerce on the water front of the Potrero, and South San Francisco, would cause a "draw" to be kept open continually. One of the San Francisco papers has suggested the expedient of removing "the heart of San Francisco" as far south as the Rolling Mill point, so that trains might be run into it; but it is doubtful if the heart of San Francisco beats responsive to the suggestion. Seriously, the whole scheme smacks so strongly of *outside property* that it is impossible to disguise it.

Nobody doubts that San Francisco must continue to be the metropolis of the Pacific coast; and the shallow efforts of a few speculators to frighten San Franciscans into committing an outrage against themselves, and their posterity, is highly reprehensible. Already, more than one-fourth the population of the State is congregated at San Francisco; and considering her established advantages, and the more luxurious habits of city folk, as compared with those of country folk, her "city trade" may be estimated at nearly one-half of the trade of the State, exclusive of the export trade. Of the California domestic trade, outside of the city, she commands, and must ever command, the lion's share. The export trade belongs, and will always belong, exclusively to San Francisco, for she supplies money for the movement of crops, and has a deeper interest than Oakland will ever have in the economical handling of our export products. If machine shops are built at Oakland, San Francisco men, with San Francisco capital, will build them; and the profits of such enterprises will return to the fountain-head.

In short, Oakland is an invaluable adjunct to the commerce of San Francisco; and far-seeing San Franciscans are proud, not jealous, of Oakland.

THE UNIVERSITY OF CALIFORNIA.

The University of California was created with the view of carrying the public educational system of the State up to its highest expression, in an institution which should realize the broadest, freest, most liberal, and most advanced ideas of University education. It receives its support from the extensive land-grants made by the General Government to the State of California, for the establishment of Agricultural and Mechanic Arts Colleges—a foundation which has been enlarged by a liberal appropriation from the State Legislature. The University, accordingly, is a State institution, and, as such, must be of equal interest to the people of every section of California. Yet, the sphere of its activity is not bounded by the lines of our own State, for its register shows that it already draws from every State and Territory of the Pacific Coast, from Mexico, from South America, and from the islands of the sea—a fact which strikingly illustrates the scope of the benefits diffused by our young but progressive University.

The Act creating the University of California was passed by the State Legislature at the session of 1867-8. It placed the supreme control of the institution in a Board of Regents which is, at present, composed of the following gentlemen of well-known culture, public spirit, and business ability:

EX-OFFICIO REGENTS.

His Excellency HENRY H. HAIGHT, Governor.
His Honor WILLIAM HOLDEN, Lieutenant-Governor.
Hon. GEORGE H. ROGERS, Speaker of the Assembly.
Hon. O. P. FITZGERALD, D.D., State Superintendent of Public Instruction.
Hon. CHARLES F. REED, President of the State Agricultural Society.
A. S. HALLIDIE, Esq., President of the Mechanics' Institute of San Francisco.

APPOINTED REGENTS.

JOHN T. DOYLE, Esq.,
Hon. RICHARD P. HAMMOND,
Hon. JOHN W. DWINELLE,
Rev. HORATIO STEBBINS, D.D.,

Hon. LAWRENCE ARCHER,
Hon. WILLIAM WATT,
Hon. SAMUEL B. MCKEE,
Hon. SAMUEL MERRITT, M.D.

HONORARY REGENTS.*

Hon. EDWARD TOMPKINS,
J. MORA MOSS, Esq.,
S. F. BUTTERWORTH, Esq.,
Hon. JOHN S. HAGER,

A. J. BOWIE, M.D.,
WILLIAM C. RALSTON, Esq.,
Hon. JOHN B. FELTON,
LOUIS SACHS, Esq.

OFFICERS OF THE BOARD OF REGENTS.

His Excellency H. H. HAIGHT, President.
ANDREW J. MOULDER, Esq., Secretary.
WILLIAM C. RALSTON, Esq., Treasurer.

The University went into operation September 23d, 1869, with Professors John and Joseph LeConte, Fisher, Swinton, Carr, Kellogg, Welcker, Pioda, Santi, and Ogilby, as the faculty. Professor John LeConte was appointed Acting-President by the Regents, and he continued in this position till the close of the scholastic year ending with July, 1870. The second year of the University began September 23d, 1870. In the intervening vacation, the Board of Regents had elected to the Presidency, Professor Henry Durant. The Register gives the following names, as composing the Faculty and Officers of the University:

HENRY DURANT, A.M., President, and Professor of Mental and Moral Philosophy.
STEPHEN J. FIELD, LL.D., Non-resident Professor of Law.
JOHN LECONTE, M.D., Professor of Physics, Industrial Mechanics, and Physiology.
JOSEPH LECONTE, M.D., Professor of Geology, Natural History, and Botany.
MARTIN KELLOGG, A.M., Professor of Ancient Languages.
General W. T. WELCKER, Professor of Mathematics.
PAUL PIODA, Professor of Modern Languages.
EZRA S. CARR, M.D., Professor of Agriculture, Chemistry, Agricultural and Applied Chemistry, and Horticulture.

---

*The term "Honorary," applied to these Regents, indicates only the mode of their election, which is made by the Ex-officio and Appointed Regents. Every Regent, however appointed, is a voting, legislative, and executive member of the Board.

WILLIAM SWINTON, A.M., Professor of the English Language and Literature, Rhetoric, Logic, and History.

THOMAS BENNETT, M.D., Professor of the Principles and Practice of Medicine.

JAMES BLAKE, M.D., Professor of Midwifery.

J. C. SHORB, M.D., Professor of Clinical Medicine.

J. D. B. STILLMAN, M.D., Professor of Materia Medica.

C. F. BUCKLEY, M.D., Professor of Anatomy.

GEORGE DAVIDSON, A.M., (Assistant U. S. Coast Survey), Non-resident Professor of Astronomy and Geodesy.

Colonel FRANK SOULÉ, Assistant Professor of Mathematics.

ROBERT E. OGILBY, Instructor in Drawing.

GEORGE TAIT, A.M., Assistant Professor of Ancient Languages.

Professor WILLIAM SWINTON, Librarian.

It is believed that the history of education in the United States presents a no more signal success, in the founding of a high institution of learning, than that which has attended the University of California. Opening with about forty students at the beginning of the first year, it has now on its catalogue the names of seven hundred and ninety members of the several Colleges and of the Preparatory Department.

The University consists of five distinct and independent Colleges, viz.: four Colleges of Arts, and one College of Letters, as follows:

1. A State College of Agriculture,
2. A State College of Mechanic Arts,   }   *Colleges of Arts.*
3. A State College of Mines,
4. A State College of Civil Engineering.
5. A State College of Letters.

The full course of instruction in each College embraces all appropriate studies, and continues for at least four years. Each College confers a proper degree, at the end of the course, upon such students as are found, upon examination, to be proficient therein.

Partial courses are organized in each of the Colleges for students "who may not desire to pursue a full course therein."

Besides the students pursuing the regular courses, any resident of California, of approved moral character, has the right to enter himself in the University as a student at large, and receive tuition in any branch or branches of instruction, at the time when the same are given in the regular course, provided his preparatory studies have been such as to qualify him to pursue the selected branches; and provided, further, he selects a sufficient number of branches — the number being designated by the Faculty.

Measures have been taken to carry out the provisions of the Act creating the University, in respect to military instruction and discipline. Acting under directions from the Board of Regents, Professor Welcker and Assistant Professor Soulé, graduates of the West Point Academy, have organized the Battalion of the University Cadets. All able-bodied male students of the University are required to attend the military exercises. The utility of such instruction and discipline is generally conceded.

The University already possesses excellent apparatus, recently procured from Europe, and valued at over $30,000, for the use of the Physical, Chemical, and other Scientific Departments. There is also a Cabinet, rich in specimens collected from various parts of the State, and the Legislature has specially provided that the ample collections of the State Geological Survey shall be devoted to the uses of the University.

By an Act of the Legislature, passed at its last session, five Scholarships were established, each of the value of three hundred dollars a year, for four years, to be competed for by candidates for the Fourth Class. It is expected and hoped that the number of scholarships will be increased by private liberality.

From the foregoing statements which we compile from the "Register," it will be seen that the University of California, in the second year of its existence, already offers ample facilities for a thorough education. It has a large and competent faculty of instruction, and costly and complete apparatus. It opens its doors, without charge, to all of both sexes who are qualified to profit by its advantages. The enlightened founders of the University of California laid its basis upon live and modern ideas of education. It is wholly free from ancient scholastic precedents and routine. It recognizes the equal dignity and worth of all knowledges and arts, and hospitably affords opportunities to students desirous of pursuing any specialty. Those who are enrolled as "students at large" can select their own studies, and attend the exercises of any of the classes. There are still shorter courses for those who can stay but a single term, or attend but a single course of lectures. If any one wishes to study some practical branch of learning—for example, metallurgy or agricultural chemistry—he will find here every facility for its prosecution. In fine, it is a *University* in the full scope and meaning of the term.

The University, while awaiting the erection of college edifices upon its extensive and beautiful domain at Berkeley, (near Oakland), is occupying the old College of California building, in this city, where it is probable the institution will remain for a considerable time to come. The striking exhibit elsewhere made of the healthfulness of Oakland, shows that in this respect it could not have been more fortunately located.

PREPARATORY DEPARTMENT OF THE UNIVERSITY.

Our sketch of the University would be far from complete, did we fail to notice the recently created training-school, or "Preparatory Department."

The necessity of some training-school which should serve as a link between the public-school system and the University, was felt soon after the latter went into operation. It was at first sought to supply this link by the organization of a Fifth Class. This was begun at the beginning of the last scholastic year, in September, 1870. The experiment was a complete success—very large numbers of pupils of both sexes having joined the "Fifth Class." Indeed, so unexpected was the increase of the class, that it was found necessary to purchase the Brayton school property, in order to afford accommodations for the students presenting themselves. In January, 1871, this class, while still retaining its distinctive name, was greatly enlarged in its scope by dividing it into various grades: thus establishing a real training-school or preparatory department. This department of the institution was put under the direction of Mr. George Tait, aided by an adequate corps of excellent teachers. We believe the department now numbers (day-scholars and boarders) upward of two hundred. It shows all the signs of enlarging and lasting usefulness. 292552

PRIVATE EDUCATIONAL INSTITUTIONS.

PACIFIC THEOLOGICAL SEMINARY.—The seminary is under the auspices of the Congregational Church. It has recently purchased the property of the Female College of the Pacific, on Academy Hill, and the regular exercises were commenced in June, 1871. Revs. George Mooar, D.D., and J. A. Benton, D.D., are Professors. There is a primary department, termed the Golden Gate Academy, and the number of students in both is about twenty-five.

MILLS SEMINARY.—Located near Fruit Vale, about four miles from Oakland. The Mills Seminary enjoys a quiet seclusion, and is yet in almost hourly communication with the metropolis. Rev. C. T. Mills is Principal, and Rev. Eli Corwin is his associate. There are two hundred young lady-students, and in all its departments the seminary is complete, and to it is conceded the position of the leading institution for the education of girls on the Pacific coast, and is by many deemed superior to any institution in the Eastern States.

OAKLAND SEMINARY AND FEMALE COLLEGE OF THE PACIFIC.—This institution has been formed by the consolidation of the Female College of the Pacific, and Mrs. Blake's Oakland Seminary; and the seminary buildings, in Oakland, on Washington Street, between Eleventh and Twelfth Streets, are occupied. The Rev. E. B. Walsworth is Principal, and he has called to his assistance an efficient corps of teachers. There are one hundred scholars.

OAKLAND MILITARY ACADEMY.—This military institution, opened January 9th, 1865, is the first of the kind that has been established on this coast. Rev. D. McClure is the proprietor and Principal. The academic staff is composed of nine experienced teachers. The buildings are situated on a rise of ground, known as Academy Hill, about a mile from the Broadway Station, and may be reached by the Telegraph Avenue cars. In the academic department, well-defined and extensive courses of study are pursued in the English branches, ancient and modern languages, natural science, mathematics, and commercial knowledge, such as will prepare students for college or business. The institution is also organized as a military post, and it is obligatory upon every student to attend the daily military drill, and perform the duties of a cadet, which do not interfere with hours of study.

LINDEN LANE BOARDING SCHOOL.—This school is located on Linden Lane, near Telegraph Avenue, about two miles from Broadway Station. The number of scholars is limited to sixteen, and the course of study is designed to fit boys to enter the university or any college. D. C. Stone, A.M., is proprietor and Principal of the school.

CONVENT OF OUR LADY OF THE SACRED HEART.—This is a girls' day and boarding-school, and is located on Webster Street, at the head of Lake Merritt. It was dedicated in the summer of 1868. The classes are taught by "Sisters of the Holy Names of Jesus and Mary," who came from Canada. The school contains sixty-one boarding-scholars and fifty-two day-scholars, and is in charge of St. Mary's Catholic Church, having been built through the efforts of Rev. Father King.

MADAME BOULLET'S SCHOOL.—Among the private schools of Oakland is a modest little establishment, at the corner of Franklin and Fifth Streets, which has been conducted for many years by Madame and Mademoiselle Boullet—Parisian ladies. The boarders are limited to ten or twelve little girls, and the number of day-scholars is also limited. Notwithstanding the unpretending character of the school, it has long been justly celebrated for the parental care exercised over the pupils, and the remarkable proficiency they acquire in the French language.

St. Joseph's Academy.—This school is for boys, and is conducted by the Christian Brothers. It is located at the corner of Jackson and Fifth Streets. Brother Gustavus is Principal, and the assistants are Brothers Alexander, Baptiste, and Thomas. It was opened July 5th, 1870, with forty-five pupils, and at the close of the December term, 1870, had eighty-five scholars in attendance.

J. C. Hyde's Day and Boarding-School.—This school is located on the corner of Harrison and Sixth Streets, and has an attendance of about twenty scholars, all boys.

The Sisters' School.—This school is located on Eighth Street, between Grove and Jefferson, and is taught by Sisters Mary Augustine and Mary Prescelle, and has an attendance of about seventy-five day-scholars, all of them girls.

Mrs. Brown's and Miss Daniels' Day-School.—This school is located on Eleventh Street, between Alice and Harrison Streets.

French and English School.—Madame D'Hierry's French and English day-school is on Seventh Street, between Grove and Castro.

Alameda Academy.—This institution was opened January 2d, 1871. Prof. J. T. Doyen is Principal.

Miss Barnes' School.—Miss Mary Barnes has a private day-school, on the corner of Sixth and Clay Streets, with an attendance of fifty pupils.

Mrs. Fogg's School.—Mrs. George H. Fogg's day-school, corner of Franklin and Second Streets, has an attendance of twelve scholars.

Brooklyn Private School.—Mrs. True has a flourishing private school in Brooklyn, with an attendance of twenty-six girls and six boys.

PUBLIC SCHOOLS.

There are 5,436 children in Alameda County between the ages of five and fifteen years, 3,269 of whom are enrolled as attendants at the public schools. There are 1,268 pupils in the public schools of Oakland. There are in the county, outside of Oakland, 66 schools, giving employment to 51 teachers. In Oakland, there are six public school buildings, giving employment to 31 regular and four special teachers. The total value of public school property in the city is $129,000. The schools now open are as follows:

High School.—Corner of Market and West Twelfth Streets. Cost of premises, $37,376 22. Principal, J. B. McChesney. Number of teachers, 3; number of scholars, 65.

Lafayette Grammar School.—Location in High School building. Principal, J. B. McChesney. Number of teachers, 8; number of scholars, 321.

Prescott Grammar School.—Second Street (West Oakland). Cost of building, $10,000. Principal, A. W. Brodt. Number of teachers, 3; number of scholars, 55.

LAFAYETTE PRIMARY.—Corner of Twelfth and Jefferson Streets. Cost of building, $17,000. Principal, Mrs. M. W. Phelps. Number of teachers, 8; number of scholars, 340.

PRIMARY No. 2.—Corner of Alice and Sixth Streets. Cost of building, $1,200. Principal, Miss F. Brigham. Number of teachers, 3; number of scholars, 125.

PRIMARY No. 3.—Corner of Grove and Fourth Streets. Cost of building, $1,200. Principal, Miss Aldrich. Number of teachers, 4; number of scholars, 201.

In addition to these, there is an evening school, taught by F. M. Campbell, City Superintendent of Public Schools, a French and a German class, which would swell the number of pupils to 1,409, and the number of teachers to 35.

CHURCHES.

FIRST CONGREGATIONAL.—Broadway, east side, between Tenth and Eleventh Streets. Organized December 9th, 1860. Temporary Pastor—George Mooar, D. D. Deacons—T. B. Bigelow, E. P. Flint, R. E. Cole, and T. L. Walker. Trustees—R. E. Cole, E. P. Flint, E. P. Sanford, Israel W. Knox, Wm. K. Rowell, and H. A. Palmer.

SECOND CONGREGATIONAL.—Oakland Point. Organized May 31st, 1868. Pastor—Rev. S. D. Gray. Trustees—Jas. A. Folger, H. G. McLean, H. C. Emmons, E. E. Walcott, and L. P. Collins.

FIRST PRESBYTERIAN.—South-east corner of Broadway and Thirteenth Streets. Organized in 1852. Pastor—D. W. Poor, D.D. Elders—Samuel Percy, Elijah Bigelow, J. J. Gardiner, Wm. C. Dodge, and G. W. Armes. Trustees—E. C. Sessions, Wm. C. Dodge, Wm. H. Miller, J. J. Gardiner, Elijah Bigelow, J. M. Selfridge, and J. Shanklin.

INDEPENDENT PRESBYTERIAN.—South-east corner of Jefferson and Twelfth Streets. Organized February 28th, 1869. Pastor—Rev. L. Hamilton. Trustees—George C. Potter (Chairman), Henry Durant, David McClure, Charles Webb Howard, J. P. Moore, John R. Glascock, J. S. Emery, N. W. Spaulding, and Hiram Tubbs. Elders—Henry Durant and David McClure. Treasurer—William B. Hardy.

MISSION CONGREGATIONAL.—Second Street, between Broadway and Washington. Organized in the summer of 1868, under the control of the First Congregational Church.

FIRST BAPTIST.—Corner of Brush and Fourteenth Streets. Organized in 1854. No permanent Pastor, at present. Deacons—William Watts and G. W. Dam. Trustees—A. L. Warner, G. W. Dam, J. F. Havens, William Watts, and A. W. Brodt. Church Clerk, A. W. Brodt; Treasurer, B. F. Pendleton.

ST. JOHN'S EPISCOPAL.—Corner of Grove and Seventh Streets. Organized June, 1852. Rector—Rev. Benjamin Akerly. Vestrymen—Rev. Benjamin Akerly (President), Gen. R. W. Kirkham (Senior Warden), Samuel Brockhurst (Junior Warden), Charles D. Haven (Secretary and Treasurer), James De Fremery, J. N. Olney, and R. H. Bennett.

St. Paul's Episcopal.—South-west corner of Webster and Twelfth Streets. Organized 1871. Rector—Rev. C. W. Turner. Vestrymen—John A. Stanley, A. I. Gladding, W. C. Parker, T. J. Hyde, Watson Webb, J. B. Harmon, R. C. Alden, Dr. Babcock. Senior Warden—A. I. Gladding. Junior Warden—Watson Webb.

St. Mary's Roman Catholic.—Seventh Street, between Grove and Jefferson. Pastor—Rev. Michael King. Assistants—Fathers Byrne and Starra.

Methodist Episcopal.—South-west corner of Washington and Ninth Streets. Pastor—Rev. T. S. Dunn. Trustees—M. T. Holcomb, J. Stratton, J. W. Carrick, James C. Stratton, and C. H. Bradley.

THE MOUNTAIN VIEW CEMETERY.

Several years ago, leading citizens of Oakland, Brooklyn, and Alameda Townships, secured a suitable location as a burial place for the dead. It comprises about two hundred acres of undulating ground at the foot-hills, about two miles eastwardly from Oakland. The Mountain View Cemetery Association was organized, and, under the operation of the State law, the ground has been dedicated forever to the sacred purposes for which it was obtained. Mr. Fred. Law Olmstead, who laid out Central Park, in New York City, was employed to survey the ground and lay out a plan for the cemetery. The plan presented by him was adopted. Improvements of a high order have already been made ; and the officers of the Association comprise gentlemen whose reputation affords a guarantee that its affairs will be attended to with a view of making the cemetery all that could be desired.

INSTITUTION FOR THE DEAF AND DUMB, AND BLIND.

The State Asylum for the education of the deaf and dumb, and the blind, is located about four miles north of Oakland, on grounds adjoining those of the University. It is one of the most beneficent of our State institutions, and is exceedingly interesting to visitors who care to see how novel and ingenious modes of instruction, and patient endeavors, are made to overcome the greatest obstacles to mental development. The building, a massive stone edifice, is considered by many to be the finest piece of architecture in the State, and is supplied with all modern improvements for the comfort and convenience of its inmates, and with all the peculiar apparatus necessary for their instruction. The total cost of buildings, grounds, etc., has been about $200,000—an expenditure which indicates the liberality and thoughtfulness of our people.

The present number of pupils is eighty-five. Fifty-nine are deaf and dumb, and twenty-six are blind. The course of study embraces most of the branches usually taught in our higher academies. Facilities are also afforded for the learning of trades. The benefits of the institution, including board, tuition, and medical attendance, are free to all deaf and dumb or blind persons, between the ages of six and twenty-one years, who may be residents of the State.

The Board of Directors consists of J. Mora Moss, President ; Chas. J. Brenham, Col. John C. Hayes, I. E. Nicholson, M.D., and Col. Harry Linden. The corps of instructors in the deaf-mute department comprises Amasa Pratt, H. B. Crandall, and Henry Frank. In the blind department, C. T. Wilkinson and M. B. Clark. The Principal is Warring Wilkinson, to whom all letters of inquiry, etc., should be addressed.

SOCIETIES AND ASSOCIATIONS.

MASONIC.

LIVE-OAK LODGE No. 61, F. and A. M.—Instituted May 4th, 1855. Officers— T. P. Wales, W. M.; Wm. H. Irwin, S. W.; Henry F. Evers, J. W.; A. J. Baber, S. D.; George E. Carleton, J. D.; Rev. Benjamin Akerly, Chaplain; J. E. Whitcher, Treasurer; James Lentell, Secretary; F. Chappellet and Franklin Warner, Stewards; S. Hirshberg, Tyler.

OAKLAND LODGE No. 188, F. and A. M.—Instituted November 4th, 1868. Officers—E. H. Pardee, W. M.; W. J. Gurnett, S. W.; W. S. Snook, J. W.; T. W. Bailey, Secretary; Myron T. Dusenbury, Treasurer.

OAKLAND CHAPTER No. 26, R. A. M.—Instituted May 5th, 1860. Officers— Benjamin Akerly, H. P.; George M. Blake, K.; T. P. Wales, S.; J. M. Miner, C. H.; S. Nolan, P. S.; Henry F. Evers, R. A. C.; Wm. H. Irwin, M. 3d V.; Ernst Janssen, M. 2d V.; Wm. D. Harwood, M. 1st V.; J. E. Whitcher, Treasurer; S. Hirshberg, Secretary; H. E. Hitchcock, Guardian.

ALAMEDA CHAPTER No. 36, R. A. M.—Instituted November 11th, 1868. Officers—N. W. Spaulding, H. P.; Walter Van Dyke, K.; E. H. Pardee, S.; C. C. Knowles, C. H.; W. J. Gurnett, P. S.

INDEPENDENT ORDER OF ODD FELLOWS.

OAKLAND LODGE No. 118.—Instituted July 3d, 1865. Officers—S. P. Knight, N. G.; R. Dalziel, V. G.; John Demott, R. S.; A. B. Brower, P. S.; Chas. Barlow, T.; Wm. L. McKay, Peter Baker, and S. K. Hassinger, Trustees.

UNIVERSITY LODGE No. 144.—Instituted June 20th, 1868. Officers—M. S. Hurd, N. G.; T. A. Bell, V. G.; C. J. Robinson, R. S.; George E. Farwell, P. S.; J. V. B. Goodrich, T.

ALAMEDA DEGREE LODGE No. 5.—Instituted February 13th, 1869. Officers—W. J. Gurnett, N. G.; J. Barnett, V. G.; S. H. Goddard, Secretary; Geo. H. Fogg, Treasurer.

GOLDEN RULE ENCAMPMENT No. 34.—Officers—J. Ingols, C. P.; S. K. Hassinger, H. P.; R. Dalziel, S. W.; B. Van Vrankin, J. W.; C. H. Townsend, Secretary; A. B. Brower, Treasurer; J. E. Marchand, J. Lufkin, and B. C. Austin, Trustees.

ODD FELLOWS' HALL ASSOCIATION.—Incorporated June, 1869. Location of building, north-west corner of Franklin and Eleventh Streets. Capital stock, $16,000. Directors—J. E. Marchand, President; T. J. Murphy, Vice-President; W. J. Gurnett, Secretary; J. L. Browne, Treasurer; W. L. McKay, Peter Baker, and O. H. Burnham.

ODD FELLOWS' LIBRARY ASSOCIATION.—Organized August 12th, 1867. Number of volumes, 2,500, free to members of contributing Lodges, of which there are two—Oakland Lodge No. 118, and University Lodge No. 144. Trustees—From Oakland Lodge, S. K. Hassinger and W. Clayton; from University Lodge, F. L. Taylor, C. J. Robinson, and B. C. Austin. Officers—C. J. Robinson, President; S. K. Hassinger, Vice-President; B. C. Austin, Recording Secretary; F. L. Taylor, Corresponding Secretary; W. Clayton, Treasurer; A. B. Brower, Librarian.

MISCELLANEOUS.

ALAMEDA STAMM NO. 113, I. O. R. M.—Organized 1867. Officers—W. Jordan, O. Ch.; Henry Kornahrens, U. Ch.; P. Ferman, R. S.; A. Koop, Treasurer; A. Eisenbach, F. Secretary.

CHEROKEE TRIBE NO. 127, IMPROVED ORDER OF RED MEN.—Organized 1869. Officers—H. Nagle, S.; A. T. Potter, S. S.; Wm. Ballantyne, G. S.; W. T. Myles, K. of V.; J. C. Plunket, C. of R.

ATHENS LODGE, I. O. G. T.—Organized 1867. Officers—G. M. Blake, W. C. T.; S. Campbell, P. W. C. T.; Miss Irwin, W. V. T.; T. Bell, W. S.; A. B. Brower, W. F. S.

TURN VEREIN.—Organized 1866. Officers—D. Vogt, President; Wm. Hummeltenberg, Vice-President; Henry Sohst, First Secretary; George Bundat, Second Secretary; H. Heyer, Treasurer; Wm. Koch, Librarian; G. Kraft, First Leader; J. Nitman, Second Leader.

OAKLAND BENEVOLENT SOCIETY.—Organized 1869. Officers—Dr. R. E. Cole, President; F. S. Page, Secretary; Dr. B. F. Pendleton, Treasurer; I. W. Knox, Rev. J. E. Benton, and G. W. Armes, Trustees.

KNIGHTS OF PYTHIAS.—Organized 1870. Officers—R. Swarbrick, V. P.; Charles A. Perkins, W. C.; D. B. Bankhead, V. C.; Wm. Parish, G.; Samuel Bailey, R. S.; Charles Parry, F. S.; F. W. Butler, B.; Wm. Myles, I. G.; E. G. Jones, O. G.

OAKLAND HEBREW BENEVOLENT SOCIETY.—Organized 1862. Officers—Jacob Letter, President; Henry Ash, Vice-President; S. Beal, Treasurer; S. Hirshberg, Secretary; N. Rosenberg, J. Alexander, L. Greenbaum, Trustees.

ST. JOSEPH'S BENEVOLENT SOCIETY.—Organized 1867. Officers—John Kearney, President; P. R. Sheehan, Vice-President; John Carry, Secretary; Patrick Scully, Treasurer; Thomas Dagnan, Clerk; Dr. S. Belden, Physician.

ANCIENT SONS OF HIBERNIA.—Organized July 7th, 1870. Officers—James McGuire, President; J. O'Connell, Vice-President; S. D. Cronin, Corresponding Secretary; John Teague, Financial Secretary; E. Fitzgerald, Treasurer. The Society numbers one hundred members.

ALAMEDA COUNTY MEDICAL ASSOCIATION.

Organized October 25, 1869. Incorporated January 9, 1871. List of members—Clinton Cushing, M.D., President; E. Trenor, M.D., Vice-President; N. E. Sherman, M.D., Treasurer; John C. Van Wyck, M.D., Librarian; H. P. Babcock, M.D., Secretary; T. H. Pinkerton, M.D., Stillman Holmes, M.D., Joseph Leconte, M.D., John Leconte, M.D., Ezra S. Carr, M.D., R. Beverly Cole, M.D., Thomas C. Hanson, M.D., Wm. Bamford, M.D., Wm. Bolton, M.D., John Van Zandt, M.D., W. R. Fox, M.D., C. S. Coleman, M.D.

MILITARY COMPANIES.

OAKLAND GUARD.— Organized in 1861. Officers — Alfred W. Burrell, Captain; John C. Orr, First Lieutenant; E. R. Turner, Second Lieutenant; H. Maloon, Orderly Sergeant.

LIVE-OAK ZOUAVES.— Organized in 1870. Officers — E. J. Kelley, Captain; Thomas Treanor, First Lieutenant; John F. Teague, Second Lieutenant; James Marchand, Orderly Sergeant.

OAKLAND GRENADIERS. — Organized in 1870. Officers — J. Callaghan, Captain; A. Herrin, First Lieutenant; S. Cronin, Second Lieutenant.

OAKLAND BANK OF SAVINGS.

Organized August 27, 1867. Capital stock, $150,000. Capital increased March 30, 1869, to $300,000; increased May 9, 1871, to $1,000,000.

OFFICERS—P. S. Wilcox, President; J. L. Browne, Cashier.

BOARD OF DIRECTORS—P. S. Wilcox, E. M. Hall, Samuel Merritt, T. B. Bigelow, Walter Blair.

The following is from the report of July 1, 1871:

```
Stock and reserve fund.........................$141,974 21
Due Depositors ................................. 246,098 01
Due Dividend No. 8............................. 22,414 25
                                              ————$410,486 47

Loans and bonds...............................$340,645 48
Office Furniture............................... 2,470 78
Stamps and currency .......................... 696 00
Cash in vault, San Francisco, and New York....... 66,674 21
                                              ————$410,486 47
```

UNION SAVINGS BANK.

Incorporated July 1, 1869, with a capital stock of $300,000, which was increased July 1, 1870, to $500,000.

OFFICERS—A. C. Henry, President; J. West Martin, Vice-President; H. A. Palmer, Cashier and Secretary.

BOARD OF DIRECTORS—A. C. Henry, J. West Martin, John C. Hays, E. Bigelow, E. A. Haines, Samuel Woods, Chas. Webb Howard, Hiram Tubbs, H. H. Haight, C. T. H. Palmer, S. Huff, W. W. Crane, Jr., R. W. Kirkham, R. S. Farrelly, A. W. Bowman, J. Mora Moss.

The following is extracted from the report of this bank, October 1, 1871:

```
Capital stock paid in...........................$450,000 00
Deposits ..................................... 271,484 43
Profit and loss............................... 10,681 90
                                              ————$732,166 33

Loans, bonds, etc............................$621,090 07
Cash on hand................................. 76,279 11
Sundries, including expense account, banking house,
    vaults, etc............................... 34,797 15
                                              ————$732,166 33
```

NEWSPAPERS.

There are three newspapers published in Oakland, as follows :

THE OAKLAND DAILY TRANSCRIPT—Issued every morning (Sundays excepted). John Scott, proprietor.

THE OAKLAND DAILY NEWS—Issued every morning (Sundays excepted). William Gagan, proprietor.

THE EVENING TERMINI—Issued every evening (Sundays excepted). By the Termini Company.

BROOKLYN.

Brooklyn is situated eastwardly from Oakland, and is bounded on two sides by the Estuary of San Antonio, as will be seen on the map. Its site is higher than that of Oakland, and is gently undulating. The bridge across the Estuary, connecting the two places, is eighty feet wide. The town government is organized as follows :

TOWN OFFICERS—H. A. Mayhew, President; A. Cannon, H. Tubbs, Charles Newton, H. Tum Suden, Trustees; J. F. Steen, Clerk and Treasurer; E. E. Webster, Assessor; O. Whipple, Marshal. School Trustees—A. W. Swett, C. C. Knowles, F. Buell.

The main street (Washington) has been graded and macadamized from the Twelfth-street bridge to Park Avenue. Many buildings have been erected during the past year. A large first-class hotel, with accommodations for three hundred persons, is almost completed, and many of the rooms are already engaged. The Contra Costa Water Company have laid their mains from Oakland, for the purpose of supplying the town with water, and hydrants for the use of the Fire Department have been placed at various points. The mains of the Oakland Gas Light Company have also been carried into the town.

The School Department is well organized and conducted.

The town has four churches—viz: one Presbyterian, Rev. Oliver Hemstreet; one Baptist, Rev. T. C. Jameson; St. Anthony's Catholic Church, under the supervision of Rev. Father King; and the Episcopal Church, Rev. Mr. Wilbur, Rector.

The absence of oaks in Brooklyn, which add so much to the charms of Oakland, is compensated, in a measure, by the picturesque scenery on every side. Its water front on the Estuary of San Antonio, with the rail tracks along the bank, gives it great prospective importance as a location for manufactures, and already there are several manufacturing establishments in successful operation.

At several places near the Estuary, overflowing artesian wells have been obtained by sinking one hundred and fifty feet.

ALAMEDA, AND THE WEBSTER-STREET BRIDGE.

The beautifully situated and rapidly growing town of Alameda, distant about two miles from Oakland, has been brought into direct communication with this city by the erection of a draw-bridge, spanning San Antonio Creek, from the foot of Webster Street. From the bridge, a macadamized road has been constructed over the marsh land, which is nearly a mile in breadth.   The progress of the town can not fail to be much accelerated by the completion of this important thoroughfare; and the advantages to be derived therefrom by the people of both places can not be too highly estimated.   .

The peninsula upon which the town is located, is about three and a half miles long by one mile wide, comprising an area of about 2,200 acres of remarkably fertile soil, ornamented by a profusion of oaks.   An abundance of excellent water is obtainable within a few feet of the surface.

Nature has made the Encinal a charming resort for people of rural tastes; and during the summer months its groves and parks are visited by thousands from San Francisco and neighboring places.   The township contains about five hundred families, most of whom own the property upon which they reside.   In Alameda, there are many delightful residences, including that of his Excellency, the Governor of California.

THE LOCAL RAILROAD AND FERRY.

The location of the road and wharf is shown on our map of Oakland, and a description of the wharf and slip may be found in the article quoted from the *Alta* (page 9).   There is a large depot on the southerly side of the slip, for the exclusive accommodation of the local passenger trains, and the facilities for the movement of passengers are excellent.   The local track is of heavy "fish-joint" iron, and runs up the wharf between the through-track and the carriage-way, with a safety-rail on each side where, otherwise, there would be a possibility of accident.

A STEAM FIRE-ENGINE.

As a guard against fire, an elegant locomotive—the "White Eagle"—with a steam-pump attachment, a tank-car, and coils of hose, is kept constantly in readiness, to fly to this point or that, with lightning speed.

THE FERRY SLIP

At San Francisco, is near the foot of Pacific street, but the improvements about it are inferior.   Provision has been made for the safety of passengers, but the arrangements for their comfort are not suggestive of the civilization of 1871.   We may assume the reason to be that the railroad company does not regard the location as a permanent one.   Public considerations suggest that the Board of State Harbor Commissioners should assign to the company a place near the foot of Market Street, with guarantees of permanency which would justify the construction of creditable improvements for the accommodation, not only of local travelers, but of the thousands who visit us from abroad.   We say the foot of Market Street, because the system of streets in San Francisco admits of no other proper location. From that point, and that alone, the street railroads could be made to radiate to every part of San Francisco, and equalize both convenience to travelers, and the benefits resulting to property.

RAILROAD AVENUE.

Returning to Oakland, we must admit that Railroad Avenue, through which the local road runs, is one of the least attractive streets in our city.   Nature has done her part, but the railroad company and the property holders have not done theirs.   There are six stations between the Bay and the Estuary, with miserable sheds at five, and not a respectable platform at one.   The street is not macadamized; only a few patches of sidewalk are made; and travelers from San Francisco, or elsewhere, are not favorably impressed with that portion of our city. Arrangements are in progress, however, to remedy these defects. The city, the railroad company, and the property owners on the avenue, are coöperating in the matter, and the sandy, unattractive, and tiresome street will soon be transformed into a beautiful boulevard.

THE BOAT AND CAR ACCOMMODATIONS

Are not surpassed on any similar line of travel. The steamer *El Capitan*, which performs the ferry service, is about one thousand tons burden, and is a stanch, powerful, and elegantly constructed boat. Moreover, the attention and forethought which insure punctuality and safety, are not wanting. The local Superintendent is accomplished in his profession, and unremitting in his watchfulness. The following statistics of travel and casualties, attest his efficiency, and demonstrate

THE SAFETY OF TRAVELING.

During the year 1870, the cars and boat made twelve trips per day, each way. The average number of passengers to each trip was one hundred and eighty, making four thousand three hundred and twenty passengers per day, or over one million and a half for the year—more than ten times the population of San Francisco. In this vast movement of passengers, *not one fatal accident occurred.* Only two persons were injured, and the Company was not accused of responsibility in either case. The Company has recently attached the "atmospheric brake" to its local trains, by means of which the engineer can stop his train almost instantly.

THE INCREASE OF TRAVEL

Is perceptible from month to month, and it is understood that the Company will soon multiply the trips. Indeed, it is quite evident that the time is not distant when crossings will be made every ten minutes; and persons seeking homesteads can safely depend upon realizing this prediction.

THE ESTUARY ROUTE,

Or "Creek Route," as it is commonly called, is used by boats and vessels carrying passengers and freight to and from Oakland, and Brooklyn. At present, three steamers, and a number of sailing craft, are plying on this route, which is open to competition. The importance of the estuary is alluded to elsewhere. Its improvement is a question of not much time; and those who rely upon seeing first-class passenger boats navigating its waters at an early day, will not be disappointed.

A RECREATIVE TRIP.

Thousands of people in San Francisco have never visited this side of the Bay, and are in unblissful ignorance of the attractions which it offers, and of the recreative and invigorating nature of the trips to and from Oakland. The street-car trips, from the business portion of San Francisco to or from any point in that city where residence property costs even *double* that of residence property in Oakland, consume more time than the trips between San Francisco and Oakland; and the monotony and discomfort of street-car travel make the former appear twice as long as the latter.

FARES.

The fares between Oakland and San Francisco are as follows: Monthly commutation tickets, $3; transient passengers, fifteen cents for regular line, except Sundays, when tickets for crossing and recrossing are sold for twenty-five cents. The fare by the opposition boat—the *Chin-du-Wan*—is ten cents.

ALAMEDA COUNTY STATISTICS.

The following report of the agricultural products, improvements, and general industries of the county, for 1870, is from the books of the County Assessor, Edwin Hunt:

AGRICULTURAL PRODUCTS.

Land inclosed, acres	91,328	Potatoes, acres	1,013
Land cultivated, acres	117,763	Potatoes, bushels	82,640
Wheat, acres	65,991	Sweet Potatoes, acres	None.
Wheat, bushels	1,017,031	Sweet Potatoes, bushels	None.
Barley, acres	36,030	Onions, acres	293
Barley, bushels	505,670	Onions, bushels	25,108
Oats, acres	3,240	Hay, acres	7,465
Oats, bushels	98,460	Hay, tons	12,475
Rye, acres	2,510	Flax, acres	375
Rye, bushels	137,000	Flax, pounds	68,600
Corn, acres	562	Hops, acres	5
Corn, bushels	13,180	Hops, pounds	1,870
Buckwheat, acres	17	Tobacco, acres	None.
Buckwheat, bushels	204	Tobacco, pounds	None.
Peas, acres	166	Beets, tons	1,295
Peas, bushels	4,038	Turnips, tons	32
Peanuts, acres	None.	Pumpkins and Squashes, tons	1,280
Peanuts, pounds	None.	Butter, pounds	75,355
Beans, acres	599	Cheese, pounds	4,218
Beans, bushels	5,975	Wool, pounds	215,775
Castor Beans, acres	None.	Honey, pounds	4,325
Castor Beans, pounds	None.		

TREES AND VINES.

Apple Trees	86,615	Olive Trees	251
Peach Trees	13,595	Prune Trees	4,120
Pear Trees	35,568	Mulberry Trees	120
Plum Trees	21,264	Almond Trees	9,249
Cherry Trees	28,788	Walnut Trees	1,552
Nectarine Trees	962	Gooseberry Bushes	43,739
Quince Trees	1,992	Raspberry Bushes	725,882
Apricot Trees	3,566	Strawberry Vines	5,758,860
Fig Trees	1,015	Grape Vines	136,148
Lemon Trees	38	Blackberry Bushes	32,200
Orange Trees	23		

WINES AND LIQUORS.

Wines, gallons	3,080	Brandy, gallons	500

LIVE STOCK.

Horses	6,525	Sheep	45,276
Mules	733	Hogs	34,772
Asses	11	Chickens	57,051
Cows	4,063	Turkeys	3,791
Calves	2,462	Geese	971
Beef Cattle	1,881	Ducks	7,042
Oxen	327	Hives of Bees	318
Total No. Cattle, incl'g Stock Cattle	16,002		

IMPROVEMENTS.

Grist Mills	7	Acres of Wheat sown in 1870	58,750
Steam Power	5	Acres of Barley sown in 1870	41,075
Run of Stone	24	Assessed value of Real Estate	$8,084,150
Water Power	2	Assessed value of Improvements	$1,532,560
Run of Stone	3	Assessed value Personal Property	$2,164,671
Barrels of Flour made	36,470	Total assessed value Property	$11,786,381
Bushels of Corn ground	21,496	Estimated total population	24,000
Railroads	4	Registered voters	4,200
Miles in length	90½	Poll-tax collected	$7,402
Land cultivated in 1870, acres	112,750